Eberhard Freitag   Hilbert Modular Forms

T0238069

Eberhard Freitag

Hilbert Modular
Forms

Springer-Verlag Berlin Heidelberg New York
London Paris Tokyo Hong Kong

Eberhard Freitag
Mathematisches Institut
Universität Heidelberg
Im Neuenheimer Feld 288

D-6900 Heidelberg
Fed. Rep. of Germany

Mathematics Subject Classification (1980): 10-XX, 32-XX

ISBN 978-3-642-08072-2

Library of Congress Cataloging-in-Publication Data.
Freitag, E. (Eberhard) Hilbert modular forms / Eberhard Freitag. p. cm.
  Includes bibliographical references.

1. Hilbert modular surfaces. I. Title. QA573.F73 1990 516.3,52--dc20 89-26258 CIP

© Springer-Verlag Berlin Heidelberg 2010
Printed in the United States of America

2141/3140-5 4 3 2 1 0 – Printed on acid-free paper

To Fred and Ursel Dieterle

Contents

Introduction

The **Hilbert modular group**

$$\Gamma_K = SL(2, \mathbf{o})$$

is the group of all 2×2 matrices of determinant 1 with coefficients in the ring $\mathbf{o}$ of integers of a **totally real number field** $K \supset \mathbf{Q}$. This group and the corresponding spaces and functions – the **Hilbert modular varieties** and **Hilbert modular forms** – have been subject of many investigations starting with the Blumenthal papers [6].

In this book we seek to develop the theory to the extent necessary for us to understand the Eilenberg - Mac Lane cohomology groups

$$H^{\bullet}(\Gamma_K, \mathbf{C}) \qquad (\Gamma_K \text{ acts trivially on } \mathbf{C}).$$

These cohomology groups are isomorphic to the singular cohomology group of the Hilbert modular variety

$$X_K = \mathbf{H}^n / \Gamma_K.$$

Here $\mathbf{H}^n$ denotes the product of n upper half-planes equipped with the natural action of Γ_K. This action being properly discontinuous, we have

$$H^{\bullet}(\Gamma_K, \mathbf{C}) = H^{\bullet}(\mathbf{H}^n / \Gamma_K, \mathbf{C})$$
$$\text{(singular cohomology)}$$

Since the Hilbert modular variety carries a natural structure as a **quasi-projective variety**, the cohomology groups inherit a **Hodge structure**, which will also be determined in the course of the book.

From the point of view of the cohomology theory of arbitrary arithmetic groups, the Hilbert modular group is nothing but a **simplified example**. It is, however, the only special case in which the cohomology can be determined explicitly; this even includes the computation of the Hodge-numbers. In contrast to the very deep and involved methods of the general theory, the case of the Hilbert modular group can be treated in an absolutely elementary manner. For these reasons the study of the Hilbert modular group is strongly justified although it should merely be considered an introduction to more general theories. Everything necessary to determine the cohomology will be developed in this book.

The principal topics discussed in this book are

1) The reduction theory (compactification of H^n/Γ_K by h "cusps", h = class number of K).
2) The elementary theory of (holomorphic) Hilbert modular forms.
3) The evaluation of the Selberg trace formula to determine the dimensions of spaces of Hilbert modular forms of weight $r > 2$. (This has been done in a very important paper by Shimizu [57], whose lines we will follow closely.)
4) We use an algebraic geometric method to come down to the border case $r = 2$ in the dimension formula. (This case has to be treated if one is not only interested in Betti but also in Hodge numbers.)
5) We need the definition of an Eisenstein series in the border case $r = 2$ where convergence is not absolute. We will achieve this in the usual way, namely by Hecke summation and **analytic continuation** of Eisenstein series. Applying the methods of Hecke and Kloosterman the analytic continuation will be obtained in an elementary way. The idea is really quite simple: Compute the Fourier coefficients and continue them!
   With this preparatory work finished, the determination of the Hilbert modular group will then be based on two papers:
6) Matsushima and Shimura [49] determined $\overset{\bullet}{H}(\Gamma, \mathbf{C})$ in the case of an irreducible discrete subgroup $\Gamma \subset SL(2,\mathbf{R})^n$ with compact quotient H^n/Γ instead of the Hilbert modular group.
7) It was Harder [26] who transferred the theory of Matsushima and Shimura to the case of the Hilbert modular group and its congruence subgroups. He showed that the cohomology splits into two parts:
   a) The **square integrable cohomology**, which can be treated like the cocompact case.
   b) The **Eisenstein cohomology**, which is due to the cusps. It is a part of the cohomology that maps injectively if one restricts the cohomology to the **boundary**.

We will also determine the **mixed Hodge structure** in the sense of Deligne [9]. This was the subject of Mr. C. Ziegler's Diplomarbeit and has been revised by him to be included as the last paragraph of this book.

Altogether the book is somewhere in between a graduate text and a research report. It can be used as an introduction to the theory of Hilbert modular forms, the Selberg trace formula, etc. There is in fact only little intersection with van de Geer's book on Hilbert modular surfaces and as both books have a different line of approach they fit together well. Several parts of the book can also be used for seminars. Therefore I have included some appendices in which the basic facts about algebraic numbers, integration, alternating differential forms and Hodge theory are described, mostly without proofs.

Finally, I would like to express my gratitude towards Mr. Holzwarth and Mr. von Schwerin who produced the TEX-manuscript and especially to Mr. Ballweg who corrected many mistakes in the original manuscript.

Chapter I. Hilbert Modular Forms

§1 Discrete Subgroups of $SL(2, \mathbf{R})$

A discrete subgroup $\Gamma \subset SL(2, \mathbf{R})$ acts discontinuously on the upper half-plane H. The parabolic elements of Γ give rise to a natural extension of H/Γ by the so-called cusp classes. We are mainly interested in the case where this extension is compact. Our basic example is $\Gamma = SL(2, \mathbf{Z})$. The method of construction is such that it can easily be generalized to the case of several variables, i.e. discrete subgroups of $SL(2, \mathbf{R})^n$ acting on the product of n upper half-planes. This will be done in the next section (§2).

It is well known that any biholomorphic mapping of the upper half-plane

$$\mathbf{H} = \{z \in \mathbf{C} \mid \operatorname{Im} z > 0\}$$

is given by $z \mapsto Mz := \frac{az+b}{cz+d}$, where $M = \begin{pmatrix} a & b \\ c & d \end{pmatrix}$ is a matrix with real coefficients and determinant 1. The set of all these matrices is the group $SL(2, \mathbf{R})$. The matrix M is uniquely determined up to its sign

$$Mz = Nz \quad \text{for all} \quad z \in \mathbf{H} \quad \Longleftrightarrow \quad M = \pm N \,.$$

We shall frequently make use of the formulae

$$M(Nz) = (MN)(z),$$

$$\begin{pmatrix} a & b \\ c & d \end{pmatrix}^{-1} = \begin{pmatrix} d & -b \\ -c & a \end{pmatrix},$$

$$dM/dz = (cz+d)^{-2},$$

$$\operatorname{Im} M(z) = (\operatorname{Im} z)/|cz+d|^2 \,.$$

The mapping

$$SL(2, \mathbf{R}) \times \mathbf{H} \to \mathbf{H}$$
$$(M, z) \mapsto Mz$$

is continuous. Here $SL(2,\mathbf{R})$ carries the natural topology induced by the Euclidean metric of $\mathbf{R}^4$.

Description of H as a Coset Space. The point $i \in \mathbf{H}$ is a fixed point of the transformation $M \in SL(2,\mathbf{R})$ if and only if $a = d$ and $b = -c$ or equivalently

$$M'M = E = \begin{pmatrix} 1 & 0 \\ 0 & 1 \end{pmatrix}.$$

So the stabilizer of i is the **special orthogonal group**

$$SO(2,\mathbf{R}) = \{M \in SL(2,\mathbf{R}) \mid M'M = E\}$$

which is obviously a **compact subgroup** of $SL(2,\mathbf{R})$. The mapping

$$SL(2,\mathbf{R}) \longrightarrow \mathbf{H}, \quad M \mapsto Mi$$

is continuous and surjective. Since

$$Mi = Ni \iff M \cdot SO(2,\mathbf{R}) = N \cdot SO(2,\mathbf{R})$$

we obtain a bijective mapping from the coset space to the upper half-plane:

$$SL(2,\mathbf{R})/SO(2,\mathbf{R}) \overset{\sim}{\longrightarrow} \mathbf{H}$$
$$M \cdot SO(2,\mathbf{R}) \longmapsto Mi.$$

If we provide the coset space with the quotient topology (a set in the coset space is open iff its inverse image under the natural projection $SL(2,\mathbf{R}) \to SL(2,\mathbf{R})/SO(2,\mathbf{R})$ is open in $SL(2,\mathbf{R})$), this mapping is continuous. But we can show even more:

1.1 Remark. *The mapping*

$$SL(2,\mathbf{R})/SO(2,\mathbf{R}) \to \mathbf{H}$$
$$M \cdot SO(2,\mathbf{R}) \mapsto Mi$$

is topological.

Proof. A bijective mapping is topological iff it is continuous and open. So we have to show that the mapping

$$SL(2,\mathbf{R}) \longrightarrow \mathbf{H}, \quad M \mapsto Mi$$

is open. It is sufficient to show that the image of a neighbourhood U of the unit matrix E is a neighbourhood of i. This is easy to be seen (it is sufficient to look at the upper triangular matrices $\begin{pmatrix} * & * \\ 0 & * \end{pmatrix} \in U$). $\square$

The description 1.1 of the upper half-plane as a coset space is compatible with the action of $SL(2,\mathbf{R})$ (the group $SL(2,\mathbf{R})$ acts on $SL(2,\mathbf{R})/SO(2,\mathbf{R})$ by multiplication from the left). An important application of 1.1 is

1.1₁ Corollary. *The mapping*

$$p: SL(2,\mathbf{R}) \to \mathbf{H}, \qquad M \mapsto Mi$$

is proper, i.e. the inverse image of a compact set is compact.

Proof. Let $K \subset \mathbf{H}$ be a compact subset. We first prove the existence of a compact subset

$$\widetilde{K} \subset SL(2,\mathbf{R}), \quad p(\widetilde{K}) \supset K.$$

For this we choose a compact neighbourhood $U(x)$ for every point $x \in SL(2,\mathbf{R})$. The image $p(U(x))$ is a neighbourhood of $p(x)$. We need only finitely many of those neighbourhoods to cover K. The union of the corresponding neighbourhoods $U(x)$ is $\widetilde{K}$. We obviously have

$$p^{-1}(K) \subset \widetilde{K} \cdot SO(2,\mathbf{R}).$$

So $p^{-1}(K)$ is compact (because it is a closed subset of $\widetilde{K} \cdot SO(2,\mathbf{R})$, which is the image of the compact set $\widetilde{K} \times SO(2,\mathbf{R})$ under a continuous mapping (multiplication)). □

1.2 Proposition. *A subgroup $\Gamma \subset SL(2,\mathbf{R})$ is discrete if and only if it acts discontinuously on $\mathbf{H}$.*

Let us recall:

a) A subset $\Gamma \subset SL(2,\mathbf{R})$ is **discrete** if the intersection of Γ with any compact subset $K \subset SL(2,\mathbf{R})$ is a finite set.

b) A subgroup $\Gamma \subset SL(2,\mathbf{R})$ **acts discontinuously** if for any two compact subsets $K_1, K_2 \subset \mathbf{H}$ the set

$$\{M \in \Gamma, \; M(K_1) \cap K_2 \neq \emptyset\}$$

is finite. It is sufficient to consider the case $K = K_1 = K_2$ because we may replace K_1, K_2 by $K_1 \cup K_2$.

Proof. 1) Assume that Γ acts discontinuously. We choose a compact neighbourhood U of the unit matrix E in $SL(2,\mathbf{R})$. We denote by V its image in $\mathbf{H}$ under the projection p. We obviously have

$$M \in U \Longrightarrow M(V) \cap V \neq \emptyset.$$

By assumption there exist only finitely many M.

2) Assume that Γ is discrete. Let $K \subset \mathbf{H}$ be a compact subset. Its inverse image $\widetilde{K} = p^{-1}(K)$ in $SL(2,\mathbf{R})$ is compact (1.1_1). We obviously have

$$M(K) \cap K \neq \emptyset \Longrightarrow M \in \widetilde{K} \cdot \widetilde{K}^{-1}.$$

The latter set is compact because it is the image of $\widetilde{K} \times \widetilde{K}$ under a continuous map $((x,y) \mapsto xy^{-1})$.   □

Fixed Points. We want to investigate the conditions under which a matrix $M \in SL(2,\mathbf{R})$ has a fixed point in the upper half-plane. The solution of the fixed point equation

$$\frac{az+b}{cz+d} = z$$

gives us

$$z = \frac{a - d + \sqrt{(a+d)^2 - 4}}{2c} \quad \text{if } c \neq 0.$$

From this simple calculation we see immediately that a transformation M different from the identity $(M \neq \pm E)$ has a fixed point in $\mathbf{H}$ if and only if $|a+d| < 2$, and in this case M has a single fixed point in $\mathbf{H}$. In general, a matrix $M \in SL(2,\mathbf{C})$ with

$$|\sigma(M)| < 2 \quad (\sigma(M) = a + d)$$

is called **elliptic**.

We summarize:

1.3 Remark. *A matrix $M \in SL(2,\mathbf{R}), M \neq \pm E$, has a fixed point in the upper half-plane if and only if it is elliptic. In this case it has a single fixed point in $\mathbf{H}$.*

A point $a \in \mathbf{H}$ is called an **elliptic fixed point** of a subgroup $\Gamma \subset SL(2,\mathbf{R})$ if the stabilizer

$$\Gamma_a = \{M \in \Gamma, \; Ma = a\}$$

contains an element different from the identity $(M \neq \pm E)$.

1.4 Remark. *The set of elliptic fixed points of a discrete subgroup $\Gamma \subset SL(2,\mathbf{R})$ is a discrete subset of $\mathbf{H}$.*

Proof. Each point $a \in \mathbf{H}$ has a compact neighbourhood $U \subset \mathbf{H}$. There are only finitely many $M \in \Gamma$ with the property $M(U) \cap U \neq \emptyset$, and so we have only finitely many elliptic fixed points in U.   □

1.5 Remark. *We assume that M is contained in a discrete subgroup Γ of $SL(2, \mathbf{R})$. Then the following three conditions are equivalent:*
a) M is elliptic or $M = \pm E$.
b) M is of finite order, i.e. $M^h = E$ for some $h \in \mathbf{N}$.
c) M has a fixed point in $\mathbf{H}$.

Proof. We already know a)⇔c) and so it is sufficient to show c)⇒b)⇒a).

c)⇒b): The stabilizer Γ_a of the fixed point a of M is a finite subgroup of the (discontinuous) group Γ and therefore each element of Γ_a is of finite order.

b)⇒a): Each matrix $M \in SL(2, \mathbf{C})$ of finite order is diagonalizable, i.e. there exists a matrix $A \in SL(2, \mathbf{C})$ with the property

$$AMA^{-1} = \begin{pmatrix} \zeta & 0 \\ 0 & \zeta^{-1} \end{pmatrix} .$$

This follows, for example, from the theory of the Jordan canonical form. The number ζ is necessarily a root of unity. If $\zeta \neq \pm 1$ we obviously have

$$|\sigma(M)| = |\zeta + \zeta^{-1}| < 2 . \qquad \square$$

Transformation of a Fixed Point into the Zero Point. The upper half-plane $\mathbf{H}$ is biholomorphically equivalent to the unit disc

$$\mathbf{E} = \{w : |w| < 1\} .$$

The biholomorphic mapping

$$\alpha : \mathbf{H} \longrightarrow \mathbf{E}$$
$$z \longmapsto (z - a)(z - \bar{a})^{-1}$$

transforms a given point $a \in \mathbf{H}$ into zero. If $\gamma : \mathbf{H} \longrightarrow \mathbf{H}$ is any biholomorphic mapping with fixed point a, then

$$\gamma_0 = \alpha \gamma \alpha^{-1} : \mathbf{E} \longrightarrow \mathbf{E}$$

is a biholomorphic mapping with fixed point 0. From the Schwartz lemma we know that each such γ_0 is of the form $\gamma_0 z = \zeta z$ where ζ is a complex number of absolute value 1. If γ_0 is of finite order, ζ is a root of unity. We know that each finite group $\mathcal{E}$ of roots of unity is cyclic (because $Z = \{a \in \mathbf{R} \mid e^{2\pi i a} \in \mathcal{E}\}$ is a discrete, hence cyclic subgroup of $\mathbf{R}$).

1.6 Remark. *Let $\Gamma \subset SL(2, \mathbf{R})$ be a discrete subgroup. The image of the stabilizer Γ_a of any point $a \in \mathbf{H}$ in the group $SL(2, \mathbf{R})/\{\pm E\}$ is a finite cyclic group.*

The Quotient Space H/Γ. Two points $z, w \in H$ are called equivalent with respect to our discrete subgroup $\Gamma \subset SL(2, \mathbf{R})$ if there exists a $M \in \Gamma$ with $Mz = w$. If we identify equivalent points we obtain the quotient space H/Γ with a natural projection

$$p : H \longrightarrow H/\Gamma .$$

We provide H/Γ with the quotient topology: A set in H/Γ is open iff its inverse image in H is open. To investigate the local structure of H/Γ we prove the following

1.7 Lemma. *a) Each point $a \in H$ has an open neighbourhood U with the following property: Two points of U are equivalent with respect to Γ iff they are equivalent with respect to Γ_a.*

b) Let (a, b) be a pair of Γ-inequivalent points of H. There exist neighbourhoods $U(a), U(b)$ such that no point of $U(a)$ is Γ-equivalent with any point of $U(b)$.

Notice. We may assume in both cases that $U(a)$ is invariant under the stabilizer Γ_a:

$$M(U(a)) = U(a) \quad \text{for all} \quad M \in \Gamma_a ,$$

because we may replace $U(a)$ by the (finite) intersection of all $M(U(a)), M \in \Gamma_a$.

Proof of 1.7.

a) If the statement is false we can find sequences

$$a_n \to a , \quad b_n \to a$$

such that a_n and b_n are equivalent $\mod \Gamma$, but inequivalent $\mod \Gamma_a$:

$$M_n(a_n) = b_n , \quad M_n \in \Gamma .$$

As Γ acts **discontinuously**, the sequence M_n belongs to a finite set. Taking subsequences we may assume that M_n is constant, $M_n = M$. Taking limits we obtain $Ma = a$, which contradicts our assumption $M \notin \Gamma_a$.

b) The proof is similar to a) and therefore we leave it to the reader.   □

1.7$_1$ Corollary. *The quotient space H/Γ is a surface.*

(A surface is a Hausdorff space which is locally homeomorphic to $\mathbf{R}^2$)

Proof. 1) Two different points of H/Γ can be separated by neighbourhoods, which follows easily from 1.7,b).

2) From 1.7,a) we conclude:
The natural projection

$$H/\Gamma_a \longrightarrow H/\Gamma$$

induces a topological mapping

$$U \overset{\sim}{\longrightarrow} V$$

of some neighbourhood U of the image of a in H/Γ_a onto some open neighbourhood of the image of a in H/Γ. For this reason the local structure of H/Γ at a is determined by Γ_a. During the considerations which led to the proof of 1.6 we constructed a finite group $\mathcal{E}$ of roots of unity with the following property: The mapping

$$z \longmapsto (z - a)(z - \overline{a})^{-1}$$

induces a homeomorphism

$$\mathsf{H}/\Gamma_a \overset{\sim}{\longrightarrow} \mathbf{E}/\mathcal{E} .$$

Let

$$\zeta = e^{2\pi i \nu/n} ; \quad (\nu, n) = 1$$

be a generator of the cyclic group $\mathcal{E}$. Obviously $\mathcal{E}$ consists of all roots of unity of order n. From this it is clear that two points w, w' of $\mathbf{E}$ are equivalent mod $\mathcal{E}$ iff $w^n = w'^n$. We obtain a unique bijective mapping α such that the following diagram commutes

$$
\begin{array}{ccccc}
q & \mathbf{E} & = & \mathbf{E} & q \\
\downarrow & \downarrow & & \downarrow & \downarrow \\
q \bmod \mathcal{E} & \mathbf{E}/\mathcal{E} & \overset{\alpha}{\longrightarrow} & \mathbf{E} & q^n .
\end{array}
$$

α is a homeomorphism because the two other arrows are continuous and open mappings.

The proof of 1.7_1 is now complete. We have shown

$$\mathsf{H}/\Gamma \sim \mathsf{H}/\Gamma_a \cong \mathbf{E}/\mathcal{E} \cong \mathbf{E} \quad .$$
$$\uparrow \text{ locally at a} \qquad\qquad\qquad \square$$

Cusps. We consider the closure of H in the Riemann sphere:

$$\overline{\mathsf{H}} = \mathsf{H} \cup \mathbf{R} \cup \{\infty\} .$$

The formula

$$z \mapsto \frac{az + b}{cz + d}$$

defines an action of $SL(2, \mathbf{R})$ on the larger space $\overline{\mathsf{H}}$. We use the usual conventions $\frac{a\kappa + b}{c\kappa + d} = \infty$ if $\kappa \in \mathbf{R}$ and $c\kappa + d = 0$ (note: $c\kappa + d = 0 \Rightarrow a\kappa + b \neq 0$)

and $\frac{a\infty+b}{c\infty+d} = \frac{a}{c}$ ($= \infty$ if $c = 0$). We are interested in the structure of the stabilizer Γ_κ of a discrete subgroup $\Gamma \subset SL(2,\mathbf{R})$ in a boundary point κ. For this purpose we choose any matrix

$$A \in SL(2,\mathbf{R}), \quad A\kappa = \infty$$

(for example $A = \begin{pmatrix} 0 & 1 \\ -1 & \kappa \end{pmatrix}$). We now consider the conjugate group $A\Gamma A^{-1}$ which is again a discrete subgroup of $SL(2,\mathbf{R})$ instead of Γ. The conjugation $M \mapsto AMA^{-1}$ obviously defines isomorphisms

$$\Gamma \overset{\sim}{\longrightarrow} A\Gamma A^{-1}$$
$$\Gamma_\kappa \overset{\sim}{\longrightarrow} (A\Gamma A^{-1})_\infty.$$

A matrix $M \in SL(2,\mathbf{R})$ fixes ∞ iff it is an upper triangular matrix. The corresponding transformation is then of the form

$$Mz = \varepsilon z + b; \quad \varepsilon > 0, b \in \mathbf{R}.$$

We are highly interested in the special case of translations

$$Mz = z + b \quad (\varepsilon = 1)$$

i.e.

$$M = \pm \begin{pmatrix} 1 & b \\ 0 & 1 \end{pmatrix}.$$

In this case we call M a **translation matrix** or simply a **translation**.

1.8 Definition. *The group Γ is said to have cusp ∞ if it contains a nontrivial ($\neq \pm E$) translation.*

1.9 Lemma. *If the discrete subgroup $\Gamma \subset SL(2,\mathbf{R})$ has cusp ∞, each element of the stabilizer Γ_∞ is a translation. Moreover the image of Γ_∞ in $SL(2,\mathbf{R})/\{\pm E\}$ is an infinite cyclic group.*

Proof. We consider the set

$$\mathbf{t} := \{a \in \mathbf{R} \mid z \mapsto z + a \text{ is contained in } \Gamma\}$$

of all real numbers $a \in \mathbf{R}$ such that

$$\begin{pmatrix} 1 & a \\ 0 & 1 \end{pmatrix} \quad \text{or} \quad -\begin{pmatrix} 1 & a \\ 0 & 1 \end{pmatrix}$$

is contained in Γ. Obviously $\mathbf{t}$ is a discrete subgroup of $\mathbf{R}$, hence cyclic: $\mathbf{t} = \mathbf{Z} \cdot a_0$. We have to show that each matrix of the form $\begin{pmatrix} \varepsilon & b \\ 0 & \varepsilon^{-1} \end{pmatrix} \in \Gamma$ is a translation, i.e. $\varepsilon^2 = 1$. This follows from the simple calculation

$$\begin{pmatrix} \varepsilon & b \\ 0 & \varepsilon^{-1} \end{pmatrix} \begin{pmatrix} 1 & a \\ 0 & 1 \end{pmatrix} \begin{pmatrix} \varepsilon & b \\ 0 & \varepsilon^{-1} \end{pmatrix}^{-1} = \begin{pmatrix} 1 & \varepsilon^2 a \\ 0 & 1 \end{pmatrix}.$$

This calculation shows that multiplication by ε^2 defines an automorphism

$$\mathbf{t} \longrightarrow \mathbf{t}, \qquad a \longmapsto \varepsilon^2 a,$$

which obviously implies $\varepsilon^2 = 1$. The second part of Lemma 1.9 is also clear, because the image of Γ_∞ in $SL(2, \mathbf{R})/\{\pm E\}$ is isomorphic to $\mathbf{t}$. $\qquad \square$

Before we give the definition of an arbitrary cusp, we notice that an upper triangular matrix $M \in SL(2, \mathbf{R})$ is a translation matrix if and only if $\sigma(M) = \pm 2$.
A matrix $M \in SL(2, \mathbf{R})$ with the property

$$\sigma(M) = \pm 2, \qquad M \neq \pm E$$

is called a **parabolic matrix**.

Notice. A parabolic matrix $M \in SL(2, \mathbf{R})$ has exactly one fixed point on the extended real axis $\mathbf{R} \cup \{\infty\}$.

1.10 Lemma. *Let $\Gamma \subset SL(2, \mathbf{R})$ be a discrete subgroup. For a boundary point $\kappa \in \mathbf{R} \cup \{\infty\}$ the following three conditions are equivalent:*
1) There exists a matrix $A \in SL(2, \mathbf{R})$, $A\kappa = \infty$, such that $A\Gamma A^{-1}$ has cusp infinity.
2) The latter condition is satisfied for each $A \in SL(2, \mathbf{R})$ with $A\kappa = \infty$.
3) There exists a parabolic element in the stabilizer Γ_κ.

The proof of lemma 1.10 is an immediate consequence of the preceding remarks and the fact that trace is invariant under conjugation.

A boundary point $\kappa \in \mathbf{R} \cup \{\infty\}$ is called a **cusp** of Γ if the conditions formulated in 1.10 are satisfied. We denote by $\mathbf{H}^*$ the union of $\mathbf{H}$ with the set of cusps of Γ,

$$\mathbf{H}^* = \mathbf{H} \cup \{\text{cusps of } \Gamma\}.$$

This set of course depends on our given discrete subgroup Γ. Let ∞ be a cusp of Γ. The stabilizer Γ_∞ only contains translations (1.9) and so it acts on the open set

$$U_C = \{z \in \mathbf{H} \mid \text{Im } z > C\}.$$

The relevance of the cusps to the structure of the quotient space $\mathbf{H}/\Gamma$ is obvious from the following

1.11 Proposition. *If ∞ is a cusp of Γ, then the natural projection*

$$U_C/\Gamma_\infty \longrightarrow H/\Gamma$$

is an open imbedding for sufficiently large $C > 0$.

("Open imbedding" means a topological mapping onto an open subset.)

Proof. The projection is obviously continuous and open. Therefore it is sufficient to show that it is injective for large C. This means:
From

$$\operatorname{Im} z > C \ , \ \operatorname{Im} Mz > C \ , \ M \in \Gamma$$

we have to deduce

$$M \in \Gamma_\infty \quad \text{(i.e. } M = \begin{pmatrix} * & * \\ 0 & * \end{pmatrix}\text{)}.$$

We first prove

1.11₁ Lemma. *Let $\Delta \subset SL(2,\mathbf{R})$ be a discrete subset with the following property:*
There are two real numbers $a, b \neq 0$, such that for all $n, m \in \mathbf{Z}$

$$M \in \Delta \Longrightarrow \begin{pmatrix} 1 & na \\ 0 & 1 \end{pmatrix} M \begin{pmatrix} 1 & mb \\ 0 & 1 \end{pmatrix} \in \Delta \,.$$

Then there exists a number $\delta > 0$ such that

$$M = \begin{pmatrix} * & * \\ c & * \end{pmatrix} \in \Delta \quad and \quad c \neq 0$$

implies

$$|c| \geq \delta \,.$$

Proof. Let

$$M_n = \begin{pmatrix} * & * \\ c_n & * \end{pmatrix} \,, \quad c_n \neq 0$$

be a sequence in Δ such that c_n converges to 0. If we multiply M_n by suitable translation matrices from the left and from the right

$$A_n M_n B_n = \begin{pmatrix} a_n & b_n \\ c_n & d_n \end{pmatrix} \,,$$

we get a new sequence in Δ with the same c_n and with the further property

$$|a_n - 1| \leq C|c_n| \,, \quad |d_n - 1| \leq C|c_n| \,.$$

Here C is a suitable constant depending on a and b. We obtain

$$|c_n b_n| = |(a_n - 1)(d_n - 1) + (a_n - 1) + (d_n - 1)|$$

$$\leq C^2 c_n^2 + 2C|c_n|.$$

Now it is obvious that the sequences a_n, b_n, c_n, d_n are bounded. By assumption Δ is a discrete subset and hence the set of all M_n is finite. □

We now deduce from 1.11_1 a lemma which obviously implies 1.11.

1.11_2 **Lemma.** *Let* $\Gamma \subset SL(2,\mathbf{R})$ *be a discrete subgroup with the cusps*

$$\infty \quad \text{and} \quad \kappa = A^{-1}\infty, \qquad A \in SL(2,\mathbf{R}).$$

Assume furthermore that a constant $C > 0$ *is given. Then there exists a constant* $C' > 0$ *such that*

$$M(A^{-1}U_C) \cap U_{C'} \neq \emptyset, \quad M \in \Gamma \qquad \Longrightarrow \qquad M\kappa = \infty.$$

Proof. The set $\Delta = \Gamma A^{-1}$ has the properties formulated in 1.11_1, because ∞ and κ are cusps of Γ. If Lemma 1.11_2 were false, we could find sequences

$$z_n \in U_C \quad \text{and} \quad M_n \in \Gamma$$

such that

$$\operatorname{Im} N_n z_n \to \infty, \quad \text{where } N_n = M_n A^{-1}$$

and

$$M_n \kappa \neq \infty \qquad \text{(i.e. } N_n \infty \neq \infty\text{)}.$$

But then we have

$$\operatorname{Im} N_n z_n = \frac{y_n}{|c_n z_n + d_n|^2} \leq \frac{1}{C|c_n|^2} \qquad \left(N_n = \begin{pmatrix} a_n & b_n \\ c_n & d_n \end{pmatrix} \right)$$

and hence

$$c_n \to 0 \quad,$$

which is a contradiction to 1.11_1. □

The structure of the quotient U_C/Γ_∞ is very easy. Let $z \mapsto z+a$, $a > 0$, be a generating translation of the stabilizer. The mapping

$$U_C \longrightarrow U_r^\bullet(0) = \{q \mid 0 < |q| < r\}, \qquad r = e^{-2\pi C/a},$$

$$z \longmapsto e^{2\pi i z/a}$$

induces a topological mapping

$$U_C/\Gamma_\infty \overset{\sim}{\longrightarrow} U_r^\bullet(0) \,.$$

Roughly speaking we may express this as follows: Close to the cusp ∞ the quotient H/Γ looks like a pointed disc $U_r^\bullet(0)$.

It looks natural to add the centre 0 of the disc to the quotient. This is done for all cusps simultaneously by means of the following construction: We introduce a topology on

$$\mathbf{H}^* = \mathbf{H} \cup \{\text{cusps of } \Gamma\}$$

(which is very different from the topology induced from the Riemann sphere). If ∞ is a cusp of Γ, the sets

$$U_C \cup \{\infty\}\,, \quad U_C = \{z \mid \operatorname{Im} z > C\}$$

will form a basis for the neighbourhoods of ∞ (and not the complements of discs as in the usual topology of the Riemann sphere).

1.12 Lemma. *The set* $\mathbf{H}^*$ *carries a unique topology with the following properties:*
a) The topology induced on $\mathbf{H}$ *is the usual one.*
b) $\mathbf{H}$ *is an open and dense subspace of* $\mathbf{H}^*$.
c) If κ *is a cusp of* Γ *and* $A \in SL(2, \mathbf{R})$ *a matrix with* $A\kappa = \infty$, *then the sets*

$$A^{-1}(U_C) \cup \{\kappa\}\,, \quad C > 0\,,$$

form a basis for the neighbourhoods of κ.

The proof of 1.12 is very easy. Of course one must know that the system of the sets $A^{-1}(U_C) \cup \{\kappa\}$ (the so-called horocycles, open discs which touch the real axis at κ together with the point κ)

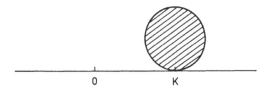

does not depend on the choice of A. The topology of $\mathbf{H}^*$ has some strange properties. We summarize some of them, the simple proofs are left to the reader.
   1) $\mathbf{H}^*$ is a Hausdorff space.
   2) The set of cusps is a discrete subset of $\mathbf{H}^*$.

3) $\mathbf{H}^*$ has a countable topology.

(Notice: The discreteness of Γ implies that Γ and therefore the set of cusps is at most countable.)

4) A cusp never has a compact neighbourhood. Otherwise the set

$$\{z \in \mathbf{C} \mid \operatorname{Im} z \geq C\} \cup \{\infty\}$$

would be a compact set for large C. But the sequence $n + Ci$, $n \in \mathbf{N}$, contains no convergent subsequence!

1.13 Proposition. *The (discrete) group Γ acts on $\mathbf{H}^*$ as a group of topological mappings. The quotient*

$$X_\Gamma = \mathbf{H}^*/\Gamma$$

(equipped with the quotient topology) is a (connected) surface, especially a locally compact Hausdorff space. The set of classes of cusps is a discrete subset of X_Γ. The canonical mapping

$$\mathbf{H}/\Gamma \hookrightarrow \mathbf{H}^*/\Gamma$$

is an open imbedding.

Remark. *For any matrix $A \in SL(2, \mathbf{R})$ the mapping $z \mapsto Az$ induces a homeomorphism (=topological mapping)*

$$X_\Gamma \overset{\sim}{\longrightarrow} X_{A\Gamma A^{-1}}$$

Proof. If κ is a cusp of Γ then $A\kappa$ is a cusp of $A\Gamma A^{-1}$. This follows immediately from the definition of a cusp and the above remark is clear from the definition of the topology of $\mathbf{H}^*$. □

We now want to investigate the structure of X_Γ close to a class of cusps. Because of the foregoing remark we restrict ourselves to the cusp ∞. From 1.11 it follows immediately that the natural projection

$$U_C \cup \{\infty\}/\Gamma_\infty \longrightarrow \mathbf{H}^*/\Gamma$$

is an open imbedding for large C. Moreover the mapping

$$z \longmapsto e^{2\pi i z/a} \quad (= 0 \text{ for } z = \infty)$$

induces a homeomorphism

$$U_C \cup \{\infty\}/\Gamma_\infty \overset{\sim}{\longrightarrow} U_r(0) = \{q \mid |q| < r\} \quad , \quad r = e^{-2\pi C/a}.$$

So, analogous to the case of inner points (1.7_1), X_Γ looks like a disc in a neighbourhood of a class of cusps. It remains to prove the **Hausdorff property**:

1) We separate the image $z_0 \in H$ from the image of a cusp, for example ∞. A simple consequence of 1.11_1 is that in each class of Γ-equivalent points in H there exists one with maximal imaginary part. Obviously

$$h(z) = \max_{M \in \Gamma}(\text{Im}(Mz))$$

depends continuously on z. The inequalities $y > C$ (including ∞) and $|z - z_0| < C^{-1}$ define open sets in H^* whose images in H^*/Γ separate the two points $[z_0]$ and $[\infty]$, if C is large enough.

2) We want to separate two different cusp classes, for example $[\infty], [\kappa]$. We choose a transformation $A \in SL(2,\mathbf{R})$, $A\kappa = \infty$. By Lemma 1.11_2 the images of $A^{-1}(U_C \cup \{\infty\})$ and $U_C \cup \{\infty\}$ in $(H)^*/\Gamma$ define disjoint neighbourhoods of the two cusp classes if C is large enough.   □

We are interested in the case where H^*/Γ is compact. In this case the number h of cusp classes is finite

$$h = \#(H^*/\Gamma - H/\Gamma).$$

Notice. Let $\Gamma_0 \subset \Gamma$ be a subgroup of finite index. Each cusp of Γ is also a cusp of Γ_0 and conversely. We therefore obtain a natural mapping

$$X_{\Gamma_0} \longrightarrow X_\Gamma,$$

which is obviously continuous. It is easy to see that this mapping is proper (the inverse image of an arbitrary compact set is compact). We therefore obtain:

X_Γ is compact if and only if X_{Γ_0} is compact.

But in general the number of cusp classes $h(\Gamma_0)$ of Γ_0 is larger than the number of cusp classes $h(\Gamma)$ of Γ.

Fundamental Sets. A subset $F \subset H$ is called a **fundamental set** of Γ if

$$H = \bigcup_{M \in \Gamma} M(F).$$

(Of course H itself is a fundamental set, but we are interested in smaller fundamental sets which reflect some of the global structure of H/Γ.)

If H/Γ is compact, we can always find a compact fundamental set by means of the following

1.14 Lemma. *Let*

$$f : X \longrightarrow Y$$

be a surjective continuous and open mapping between locally compact spaces.
*If $K \subset Y$ is a compact subset of Y we can find a compact subset $\widetilde{K} \subset X$
with the property*

$$f(\widetilde{K}) \supset K \, .$$

Corollary. *The discrete subgroup $\Gamma \subset SL(2,\mathbf{R})$ has a compact fundamental
set if and only if $\mathbf{H}/\Gamma$ is compact.*

The proof of 1.14 is easy and can be omitted (compare proof of 1.1_1).

Cusp Sectors. For two positive numbers s and t we define the domain

$$V(s,t) = \{ z \in \mathbf{H} \mid |x| \le s \, , \, y \ge t \} \, .$$

Let κ be a cusp of Γ and $A \in SL(2,\mathbf{R})$ a transformation which carries
κ to infinity

$$A\kappa = \infty \, .$$

We call the domain

$$A^{-1}(V(s,t))$$

a cusp sector with respect to κ. This notion is obviously independent of the
choice of A.

1.15 Proposition. *Let $\Gamma \subset SL(2,\mathbf{R})$ be a discrete subgroup such that $\mathbf{H}^*/\Gamma$
is compact. Let*

$$\kappa_1, \dots, \kappa_h$$

*be a set of representatives of the Γ-classes of cusps. Then there exists a
fundamental set*

$$F = K \cup V_1 \cup \dots \cup V_h$$

*where K is a compact subset of $\mathbf{H}$ and V_j is a cusp sector with respect to
κ_j $(1 \le j \le h)$.*

Fundamental Domains. Fundamental domains are fundamental sets which
are minimal in a certain sense and which have reasonable geometric properties. For our purposes the following definition is sufficient.

1.16 Definition. *A fundamental set $F \subset \mathsf{H}$ of Γ is called a fundamental domain if the following properties are satisfied:*
a) F is measurable.
b) There exists a set $S \subset F$ of measure 0 such that two different points of $F - S$ are inequivalent $\bmod \Gamma$.

Of course one has to clarify what "measurable" means. In our context it is sufficient to use the usual Lebesgue measure.

Example: The famous fundamental domain of $\Gamma = SL(2, \mathbf{Z})$ is defined by the inequalities

$$|z| \geq 1, \quad |\mathrm{Re}\, z| \leq \frac{1}{2}.$$

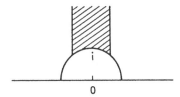

For S one can take the boundary of F. The existence of reasonable fundamental domains follows from 1.15 and the remark below, which is an immediate consequence of general facts about measurable equivalence relations. (A II.10).

1.17 Remark. *Each measurable fundamental set contains a fundamental domain.*

Final Remark. For discrete subgroups of $SL(2, \mathbf{R})$ there is quite a simple construction for a very nice fundamental domain (so-called "normal polygons" in non-Euclidean geometry). But the method described above carries more easily to the case of several variables.

§2 Discrete Subgroups of $SL(2, \mathbf{R})^n$

We will generalize the constructions of §1 to the case of discrete subgroups $\Gamma \subset SL(2, I\!\!R)^n$ and describe the extension of H^n / Γ by cusps. This construction will be justified by the fact that in the case of the Hilbert modular group the extended space is compact.

We want to generalize the basic constructions of §1 to the case of several variables.

The group $SL(2, \mathbf{R})^n$ acts on the product of n upper half-planes:

$$Mz := (M_1 z_1, \ldots, M_n z_n),$$

where

$$M = (M_1, \ldots, M_n), \quad z = (z_1, \ldots, z_n).$$

We shall also use the notation

$$M = \begin{pmatrix} a & b \\ c & d \end{pmatrix}; \quad a = (a_1, \ldots, a_n), \ldots$$

if

$$M_j = \begin{pmatrix} a_j & b_j \\ c_j & d_j \end{pmatrix}, \quad 1 \leq j \leq n$$

and

$$Mz = (az + b)(cz + d)^{-1}.$$

We shall occasionally use the notation

$$Sz = z_1 + \ldots + z_n$$
$$Nz = z_1 \cdot \ldots \cdot z_n$$

for $z \in \mathbf{C}^n$. Many proofs of the one-variable case (§1) carry over immediately to the case of several variables. In these cases we omit the proof. A good example for this is

2.1 Proposition. *A subgroup $\Gamma \subset SL(2,\mathbf{R})^n$ is discrete if and only if it acts discontinuously on $\mathbf{H}^n$.*

Recall: "To act discontinuously" means that for each compact set $K \subset \mathbf{H}^n$ the set

$$\{M \in \Gamma \mid M(K) \cap K \neq \emptyset\}$$

is finite.

Proof. Compare 1.2. □

We want to introduce the notion of a cusp. For this we extend the action of $SL(2,\mathbf{R})^n$ to $\overline{\mathbf{H}}^n$ where

$$\overline{\mathbf{H}} = \mathbf{H} \cup \overline{\mathbf{R}}, \quad \overline{\mathbf{R}} = \mathbf{R} \cup \{\infty\}.$$

The Cusp $\infty = (\infty, \ldots, \infty)$. We assume that a discrete subgroup $\Gamma \subset SL(2,\mathbf{R})^n$ is given and we are going to define what it means that Γ has cusp infinity. The justification of the following definition is the fact that our main example, the Hilbert modular group, has cusp ∞ and moreover that the quotient

$$\mathbf{H}^n/\Gamma, \quad \Gamma \text{ Hilbert modular group},$$

can be compactified by adding a finite number of cusps (one of them is the
cusp $\infty = (\infty, \ldots, \infty)$). The condition that ∞ is a cusp of Γ means in some
sense that the stabilizer

$$\Gamma_\infty = \{M = \begin{pmatrix} a & b \\ 0 & d \end{pmatrix} \in \Gamma\}$$

is as large as possible. We first look at the translations in Γ_∞:

Put

$$\mathbf{t} = \{a \in \mathbf{R}^n \mid z \mapsto z + a \text{ lies in } \Gamma\}.$$

Here "$z \mapsto z + a$ lies in Γ"means of course that there is an element
$M \in \Gamma$ with $Mz = z + a$. There are 2^n possibilities for M,

$$M = \left(\pm \begin{pmatrix} 1 & a_1 \\ 0 & 1 \end{pmatrix}, \ldots, \pm \begin{pmatrix} 1 & a_n \\ 0 & 1 \end{pmatrix} \right).$$

$\mathbf{t}$ is a discrete subgroup of $\mathbf{R}^n$. It is well-known that each discrete subgroup $\mathbf{t}$ of $\mathbf{R}^n$ is isomorphic to $\mathbf{Z}^k$ for some $k, 1 \le k \le n$. More precisely:
There exist k $\mathbf{R}$-linearly independent vectors

$$a_1, \ldots, a_k \in \mathbf{R}^n$$

such that

$$\mathbf{t} = \mathbf{Z}a_1 + \ldots + \mathbf{Z}a_k.$$

In the case $k = n$, $\mathbf{t}$ is called a **lattice**. So $\mathbf{t}$ is a lattice iff there exists a
basis $a_1, \ldots, a_n$ of the vector space $\mathbf{R}^n$ such that

$$\mathbf{t} = \mathbf{Z}a_1 + \ldots + \mathbf{Z}a_n.$$

Of course the basis $a_1, \ldots, a_n$ is not uniquely determined. We call the set

$$P = \{x \in \mathbf{R}^n \mid x = \sum_{j=1}^{n} t_j a_j \quad 0 \le t_j \le 1\}$$

a **fundamental parallelotope** of $\mathbf{t}$. It is a **compact** set with the property that
its translates $a + P$, $a \in \mathbf{t}$, cover the whole $\mathbf{R}^n$.

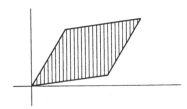

The definition of "cusp ∞" involves two conditions

2.2 First Condition. *The translation module* $\mathbf{t}$ *is a lattice.*

A vector $\varepsilon \in \mathbf{R}^n$ is called **totally positive** $(\varepsilon > 0)$ iff

$$\varepsilon_1 > 0, \ldots, \varepsilon_n > 0 \,.$$

We call a totally positive element $\varepsilon \in \mathbf{R}^n$ a **multiplier** of Γ if there exists a vector $b \in \mathbf{R}^n$ such that the transformation $z \mapsto \varepsilon z + b$ lies in Γ.
The set Λ of all multipliers is a subgroup of the (multiplicative) group of all totally positive vectors. We denote the multiplicative group of all positive real numbers by

$$\mathbf{R}_+ = \{t \in \mathbf{R} \mid t > 0\} \,.$$

2.3 Remark. *If the first condition (2.2) is satisfied, the group of multipliers Λ is a discrete subgroup of $(\mathbf{R}_+)^n$. Each multiplier satisfies*

$$\varepsilon_1 \cdot \ldots \cdot \varepsilon_n = 1 \,.$$

Proof. If ε is a multiplier, a transformation of the form $z \mapsto \varepsilon z + b$ is contained in Γ. We may replace b by $b + a, a \in \mathfrak{t}$, and therefore assume that b is contained in a certain bounded set. The discreteness of Γ now implies the discreteness of Λ. To prove the second statement of 2.3 we note that Λ acts on $\mathfrak{t}$:

$$\Lambda \times \mathfrak{t} \longrightarrow \mathfrak{t}$$
$$(\varepsilon, a) \longmapsto \varepsilon \cdot a = (\varepsilon_1 a_1, \ldots, \varepsilon_n a_n) \,.$$

Multiplication with ε defines a linear mapping

$$m_\varepsilon : \mathbf{R}^n \longrightarrow \mathbf{R}^n$$
$$a \longmapsto \varepsilon a$$

whose determinant is $\varepsilon_1 \cdot \ldots \cdot \varepsilon_n$. The matrix of m_ε with respect to a basis of the lattice is integral. Therefore we have

$$\det m_\varepsilon \in \mathbf{Z} \,.$$

The same applies to ε^{-1} instead of ε, and we obtain $\det m_\varepsilon = \pm 1$. $\square$

The multiplicative group of positive real numbers is topologically isomorphic to the additive group of all real numbers. An isomorphism is given by

$$\log : \mathbf{R}_+ \overset{\sim}{\longrightarrow} \mathbf{R} \,.$$

By means of this isomorphism we transform the group of multipliers into a discrete subgroup of $\mathbf{R}^n$:

$$\log \Lambda \subset \mathbf{R}^n$$
$$\log \varepsilon := (\log \varepsilon_1, \ldots, \log \varepsilon_n) \,.$$

If our first condition 2.2 is satisfied, $\log \Lambda$ cannot be a lattice because the condition $\varepsilon_1 \cdot \ldots \cdot \varepsilon_n = 1$ implies that $\log \Lambda$ is contained in the hyperplane (=subspace of dimension $n - 1$)

$$V = \{a \in \mathbf{R}^n \mid a_1 + \ldots + a_n = 0\} \,.$$

We obtain that

$$\Lambda \cong \mathbf{Z}^k \,, \quad k \le n - 1 \,.$$

2.4 Second Condition.

$$\Lambda \cong \mathbf{Z}^{n-1} \,.$$

2.5 Definition. *The discrete subgroup $\Gamma \subset SL(2,\mathbf{R})^n$ has cusp $\infty = (\infty, \ldots, \infty)$ if the first and the second condition (2.2 and 2.4) are satisfied, i.e.*

$$\mathbf{t} \cong \mathbf{Z}^n \,, \quad \Lambda \cong \mathbf{Z}^{n-1} \,.$$

We now give the definition of an arbitrary cusp

$$\kappa = (\kappa_1, \ldots, \kappa_n) \in \overline{\mathbf{R}}^n \,.$$

We can always find a transformation $A \in SL(2,\mathbf{R})^n$ which transforms κ to infinity, $A\kappa = \infty$.

2.6 Remark. *Let $\Gamma \subset SL(2,\mathbf{R})^n$ be a discrete subgroup and $A \in SL(2,\mathbf{R})^n$ be an element such that $A\Gamma A^{-1}$ has cusp infinity. Then for each $B \in SL(2,\mathbf{R})^n$ with*

$$A^{-1}(\infty) = B^{-1}(\infty)$$

the group $B\Gamma B^{-1}$ has also cusp infinity.

Proof. It is sufficient to treat the case

$$A = (E, \ldots, E), \quad B(\infty) = \infty \,,$$

i.e. $B(z) = \varepsilon z + b$.
Using an obvious notation we have

$$\mathbf{t}(B\Gamma B^{-1}) = \varepsilon^2 \, \mathbf{t}(\Gamma) \,,$$
$$\Lambda(B\Gamma B^{-1}) = \Lambda(\Gamma) \,,$$

which gives our assertion.   □

The preceding remark 2.6 justifies the following

2.7 Definition. *A discrete subgroup $\Gamma \subset SL(2,\mathbf{R})^n$ has cusp $\kappa \in \overline{\mathbf{R}}^n$ iff for some (every)*

$$A \in SL(2,\mathbf{R})^n, \quad A\kappa = \infty,$$

the group $A\Gamma A^{-1}$ has cusp ∞.

We again use the notation

$$(\mathsf{H}^n)^* = \mathsf{H}^n \cup \text{ set of cusps of } \Gamma .$$

2.8 Lemma. *The set $(\mathsf{H}^n)^*$ (which depends on Γ) carries a unique topology with the following properties:*

a) The topology induced on H^n is the usual one.

b) H^n is an open and dense subset of $(\mathsf{H}^n)^*$.

c) If κ is a cusp of Γ and $A \in SL(2,\mathbf{R})^n$ a transformation with $A\kappa = \infty$, then the sets

$$A^{-1}(U_C) \cup \{\kappa\}, \quad C > 0,$$

with

$$U_C = \{z \in \mathsf{H}^n \mid \prod_{j=1}^{n} \operatorname{Im} z_j > C\}$$

form a basis for the neighbourhoods of κ.

The proof is the same as that of 1.12, so we omit it. We only mention two important facts:
a) The system of sets

$$A^{-1}(U_C) \cup \{\kappa\}, \quad C > 0,$$

does not depend on the choice of A.
b) The stabilizer Γ_∞ acts on U_C, because each transformation $M \in \Gamma_\infty$ is of the form

$$Mz = \varepsilon z + a$$

with

$$N\varepsilon := \varepsilon_1 \cdot \ldots \cdot \varepsilon_n = 1 .$$

We also see immediately: *If κ is a cusp of Γ, then $A\kappa$ is a cusp of $A\Gamma A^{-1}$ for every $A \in SL(2,\mathbf{R})^n$.*

We obtain especially that Γ acts on $(\mathsf{H}^n)^*$ and therefore we can consider the **quotient space**

$$X_\Gamma = (\mathsf{H}^n)^*/\Gamma,$$

equipped with the quotient topology.

If A is any element of $SL(2,\mathbf{R})^n$, then the transformation $z \longmapsto Az$ induces a homeomorphism

$$X_\Gamma \overset{\sim}{\longrightarrow} X_{A\Gamma A^{-1}} .$$

Therefore we may reduce the investigation of the local structure of X_Γ (for arbitrary Γ) at a cusp κ to the case $\kappa = \infty$.

2.9 Proposition. *For any discrete subgroup $\Gamma \subset SL(2,\mathbf{R})^n$ the quotient space*

$$X_\Gamma = (\mathbf{H}^n)^*/\Gamma$$

is a locally compact Hausdorff space. If ∞ is a cusp of Γ, then the canonical mapping

$$U_C \cup \{\infty\}/\Gamma_\infty \longrightarrow (\mathbf{H}^n)^*/\Gamma$$

is an open imbedding for sufficiently large C. This system defines a basis for the neighbourhoods of the class of ∞.

For the proof of 2.9 one needs a suitable generalization of 1.11_1.

2.9_1 Lemma. *Let $\Delta \subset SL(2,\mathbf{R})^n$ be a discrete subset and let*

$$\Gamma, \Gamma' \subset SL(2,\mathbf{R})^n$$

be discrete subgroups with cusp ∞. We assume

$$\Gamma_\infty \cdot \Delta \cdot \Gamma'_\infty \subset \Delta .$$

Then there exists a number $\delta > 0$ with the following property:

$$\begin{pmatrix} a & b \\ c & d \end{pmatrix} \in \Delta \quad and \quad c \neq 0$$

implies

$$|Nc| \geq \delta .$$

As a special case we obtain

$$Nc = 0 \Longrightarrow c = 0 .$$

Proof. We first prove the last statement. Assume the existence of a

$$\begin{pmatrix} a & b \\ c & d \end{pmatrix} \in \Delta , \quad c \neq 0 , \quad c_1 = 0 .$$

We choose a sequence of $\alpha \in \mathfrak{t}(\Gamma'_\infty)$ such that

$$\alpha \neq 0, \ \alpha_2 \to 0, \ldots, \alpha_n \to 0.$$

Such a sequence exists by AI.19,2). We notice that by AI.19,1) all components of α are different from 0. We now consider

$$\mathrm{Im} \begin{pmatrix} a & b \\ c & d \end{pmatrix} \begin{pmatrix} 1 & \alpha \\ 0 & 1 \end{pmatrix} (i) = N\frac{1}{c^2 + (d + c\alpha)^2} \to N\frac{1}{c^2 + d^2}$$

After that we may choose a sequence of $\beta \in \mathfrak{t}(\Gamma_\infty)$ such that the real part of

$$N(i), \quad N = \begin{pmatrix} 1 & \beta \\ 0 & 1 \end{pmatrix} \begin{pmatrix} a & b \\ c & d \end{pmatrix} \begin{pmatrix} 1 & \alpha \\ 0 & 1 \end{pmatrix} = \begin{pmatrix} * & * \\ * & d + c\alpha \end{pmatrix}$$

remains bounded. We obtain a contradiction to the discreteness of Δ (since the sequence $N(i)$, hence N, because of 1.1_1 , is contained in a compact subset) . We now come to the proof of the first part of 2.9_1:
Assume that there exists a sequence

$$\begin{pmatrix} a & b \\ c & d \end{pmatrix} \in \Delta, \quad c \neq 0, \quad Nc \to 0.$$

After multiplication with suitable matrices of Γ_∞ from the left we may assume

$$\delta \leq |c_j| \ / \ \sqrt[n]{Nc} \leq \delta^{-1},$$

where δ is a suitable constant. This follows from the fact that $\log \Lambda$ is a lattice in the trace-zero hyperplane of $\mathbf{R}^n$. But now we have

$$c_j \to 0 \quad \text{for} \quad 1 \leq j \leq n.$$

Now the proof can be completed in exactly the same manner as in the case n=1 (1.11_1).   □

An immediate consequence of 2.9_1 is (compare 1.11_2)

2.9_2 Lemma. *Let* κ, κ' *be cusps of the discrete subgroup* $\Gamma \subset SL(2,\mathbf{R})^n$. *There exist neighbourhoods* U, U' *in* $(\mathbf{H}^n)^*$ *such that*

$$M(U) \cap U' \neq \emptyset, \quad M \in \Gamma \quad \Longrightarrow \quad M\kappa = \kappa'.$$

Proof of 2.9. From 2.9_2 we may deduce as in the case $n = 1$:
a) X_Γ is a Hausdorff space.
b) The mapping

$$U_C \cup \{\infty\}/\Gamma_\infty \longrightarrow (\mathbf{H}^n)^*/\Gamma$$

is an open imbedding for sufficiently large C (if ∞ is a cusp of Γ).
What remains to be shown is, that X_Γ is locally compact. Of course H^n/Γ
is locally compact because H^n is so and the projection $\mathsf{H}^n \longrightarrow \mathsf{H}^n/\Gamma$ is
open and surjective. So we only have to show that each class of cusps has a
compact neighbourhood. We actually show

2.10 Lemma. *The space*

$$\overline{U}_C \cup \{\infty\}/\Gamma_\infty$$

with

$$\overline{U}_C = \{z \in \mathsf{H}^n \mid Ny \geq C\}$$

is compact for $C > 0$.

The group Γ_∞ of course acts on the space $\overline{U}_C \cup \{\infty\}$ which carries the
topology induced from $(\mathsf{H}^n)^*$. The quotient is provided with the quotient
topology.

Proof. We first construct a fundamental set for Γ_∞.
A subset P of an n-dimensional real vector space is called a parallelotope if
there exists a basis $a_1, \ldots, a_n$ with the property

$$P = \{a \mid a = \sum_{j=1}^{n} t_j a_j , \ 0 \leq t_j \leq 1 \text{ for } 1 \leq j \leq n\} .$$

A parallelotope is of course compact.

2.10₁ Lemma. *Let $\Gamma \subset SL(2, \mathsf{R})^n$ be a discrete subgroup with cusp infinity.
There exist parallelotopes*

$$
\begin{array}{ll}
a) & P \subset \mathsf{R}^n \\
b) & Q \subset \{v \in \mathsf{R}^n \mid Sv = 0\}
\end{array}
$$

such that the domain

$$V = \{z \in \overline{U}_C \mid x \in P , \ \log \frac{y}{\sqrt[n]{Ny}} \in Q\}$$

is a fundamental set of Γ_∞ acting on $\overline{U}_C$.

Proof. Take for P a fundamental parallelotope of t in R^n and for Q a
fundamental parallelotope of $\log \Lambda$ in the vector space

$$\{v \in \mathsf{R}^n \mid Sv = 0\} .$$

Notice that

$$u = \log \frac{y}{\sqrt[n]{Ny}}$$

is contained in this vector space. A transformation of the type $z \mapsto \varepsilon z + \beta$ has the effect $u \mapsto u + \log \varepsilon$. A transformation $z \mapsto z + \alpha$ leaves u unchanged and has the effect $x \mapsto x + \alpha$. Now the proof of Lemma 2.10_1 is clear. $\square$

We now consider the closed, hence compact interval $[\sqrt[n]{C}, \infty]$ in the extended real axis. The image of the following, obviously continuous mapping

$$[\sqrt[n]{C}, \infty] \times P \times Q \longrightarrow \overline{U}_C \cup \{\infty\}$$
$$(t, x, \log y) \longmapsto \begin{cases} x + ity & \text{if } t < \infty \\ \infty & \text{if } t = \infty \end{cases}$$

is a compact fundamental set of Γ_∞ acting on $\overline{U}_C \cup \{\infty\}$. Therefore $\overline{U}_C \cup \{\infty\}/\Gamma_\infty$ is compact. $\square$

Cusp Sectors. Consider any parallelotopes

$$P \subset \mathbf{R}^n,$$
$$Q \subset \{v \in \mathbf{R}^n \mid Sv = 0\}$$

and a positive $C > 0$. The domain

$$V = \{z \in U_C \mid x \in P, \ \log \frac{y}{\sqrt[n]{Ny}} \in Q\}$$

is called a cusp sector at ∞. If κ is an arbitrary element of $\overline{\mathbf{R}}^n$,

$$A\kappa = \infty, \quad A \in SL(2,\mathbf{R})^n,$$

we call a domain of the type $A^{-1}(V)$ a cusp sector at κ. This notion is independent of the choice of A.

An obvious generalization of 1.15 is

2.11 Proposition. *Let* $\Gamma \subset SL(2,\mathbf{R})^n$ *be a discrete subgroup such that the space*

$$X_\Gamma := (\mathbf{H}^n)^*/\Gamma$$

is compact. Let be $\kappa_1, \ldots, \kappa_h$ *be a set of representatives of the (finite !) number of cusp classes of* Γ. *There exists a fundamental set*

$$F = K \cup V_1 \cup \ldots \cup V_h,$$

where K *is a compact subset of* $\mathbf{H}^n$ *and* V_j *is a cusp sector at* κ_j *$(1 \le j \le h)$.*

Singularities. In contrast to the one-variable case the cusp classes are never smooth points of X_Γ if $n > 1$. (A point is called smooth if it has an open neighbourhood homeomorphic to an open ball in $\mathbf{R}^n$).
We want to indicate the reason for readers who are familiar with homology groups.

The local homology group of a topological space X at a point $a \in X$ is defined as

$$H(X, \{a\}) := \varprojlim_{U \ni a} H_1(U - \{a\}, \mathbf{Z}),$$

where U runs over the system of neighbourhoods of a. It is enough to take a fundamental system. If $n > 2$ the space $\mathbf{R}^n - \{0\}$ is simply connected. It follows that

$$H(X, \{a\}) = 0$$

if a is a smooth point of X with dimension $X > 2$. On the other hand the fundamental group of U_C/Γ_∞ is Γ_∞ itself because Γ_∞ acts freely on the contractible space U_C. The first homology group is the abelianization

$$\Gamma_\infty^{ab} = \Gamma_\infty/[\Gamma_\infty, \Gamma_\infty]$$

where $[\Gamma_\infty, \Gamma_\infty]$ denotes the commutator subgroup of Γ_∞. We obtain

$$H(X_\Gamma, [\kappa]) \cong \Gamma_\kappa^{ab}$$

for each cusp κ.
The commutator subgroup of Γ_∞ contains only translations! We obtain that the abelianization Γ_∞^{ab} is an infinite group in the case $n > 1$. It is not hard to show that

$$\Gamma_\infty^{ab} \cong \mathbf{Z}^{n-1} \oplus \text{finite group}.$$

Elliptic Fixed Points. An immediate consequence of 1.3 and 1.5 is

2.12 Lemma. *We assume that M is contained in a discrete subgroup Γ of $SL(2, \mathbf{R})^n$. Then the following three conditions are equivalent:*
a) All components $M_1, \ldots, M_n$ of M are elliptic.
b) M is of finite order.
c) M has a fixed point in $\mathbf{H}^n$.

The fixed point of such an M is not uniquely determined even if M is non-trivial. Some of the components of M could be trivial, others not. But in the case of Hilbert's modular group the following condition will be satisfied:

2.13 First condition of irreducibility. *The restriction of each of the n projections*

$$p_j : SL(2,\mathbf{R})^n \longrightarrow SL(2,\mathbf{R}) \quad (1 \le j \le n)$$

to Γ is injective.

2.14 Remark. *Let $\Gamma \subset SL(2,\mathbf{R})^n$ be a discrete subgroup which satisfies the first condition of irreducibility. The image of the stabilizer Γ_a of any point $a \in \mathbf{H}^n$ in the group $(SL(2,\mathbf{R})/\{\pm E\})^n$ is a finite cyclic group. The set of elliptic fixed points of Γ (=set of points where this image is non-trivial) is a discrete subset of $(\mathbf{H}^n)^*$.*

Proof. Compare 1.4 and 1.6 and make use of 2.8 (notice that Γ_∞ contains no non-trivial elements of finite order). □

2.14₁ Corollary. *If (in addition to the irreducibility condition) the quotient space*

$$X_\Gamma = (\mathbf{H}^n)^*/\Gamma$$

is compact, there exists only a finite number of classes of elliptic fixed points (as well as of cusps).

We now determine the local structure of $\mathbf{H}^n/\Gamma$ (compare with the proof of 1.7₁).

Let a be any point of $\mathbf{H}^n$. By means of the transformation

$$z \longmapsto (z-a)(z-\bar{a})^{-1}$$

we may transform $\mathbf{H}^n$ to $\mathbf{E}^n$ and a to 0. The group of transformations Γ_a corresponds to a finite cyclic group G of transformations of $\mathbf{E}^n$ onto itself. A generator of G has the form

$$(w_1,\ldots,w_n) \longmapsto (\zeta_1 w_1,\ldots,\zeta_n w_n),$$

where $\zeta_1,\ldots,\zeta_n$ are roots of unity of order $e = \#G$. We assume that the irreducibility condition 2.13 is satisfied. We may assume

$$\zeta_1 = e^{2\pi i/e} ,$$

because we can replace the generator by a power coprime to e, and we then have

$$\zeta_j = e^{2\pi i e_j/e} , \quad (e_j,e) = 1 , \ 1 \le e_j \le e \quad (1 \le j \le n) .$$

The group G is completely determined by the system of numbers

$$(e_1, e_2, \ldots, e_n) \quad (e_1 = 1) .$$

We call G of type $(e_1,\ldots,e_n)$. As in the case $n = 1$ we now obtain:

2.15 Lemma. *Let* $\Gamma \subset SL(2, \mathbf{R})^n$ *be a discrete subgroup which satisfies the irreducibility condition 2.13. Let a be any point of* $\mathbf{H}^n$. *The natural projection*

$$\mathbf{H}^n/\Gamma_a \longrightarrow \mathbf{H}^n/\Gamma$$

is a local homeomorphism at a. *The transformation* $z \mapsto (z-a)(z-\bar{a})^{-1}$ *induces a homeomorphism*

$$\mathbf{H}^n/\Gamma_a \overset{\sim}{\longrightarrow} \mathbf{E}^n/G \,,$$

where G *is a finite cyclic group of order* e *and a certain type*

$$(e_1, \ldots, e_n) \qquad (e_1 = 1 \,, \; (e_j, e) = 1 \,, \; 1 \le e_j \le e \text{ for } 1 \le j \le n) \,.$$

But in contrast to the one-variable case the image of 0 in $\mathbf{E}^n/G$ is never a smooth point if $n > 1$ and of course $e > 1$. (A similar consideration as in the case of cusps shows that in the case $n > 1$ (where $\mathbf{E}^n - \{0\}$ is simply connected) the local homology group of $\mathbf{E}^n/G$ at the image of 0 is G itself.)

We may summarize:
Let $\Gamma \subset SL(2, \mathbf{R})^n$ be a discrete subgroup satisfying the irreducibility condition 2.13 and such that $X_\Gamma = (\mathbf{H}^n)^*/\Gamma$ is compact. The space X_Γ contains a finite number of singularities (non-smooth points), namely the cusp classes and the classes of the elliptic fixed points. The complement

$$X_\Gamma - \text{classes of cusps and of elliptic fixed points}$$

is a manifold of (real) dimension $2n$ (i.e. is locally homeomorphic to $\mathbf{R}^{2n}$).

The results of this section, especially the central Lemma 2.9_1 are essentially due to H. Maaß [47].

§3 The Hilbert Modular Group

We have to use some basic facts about algebraic numbers. For the convenience of the reader we have summarized them (without proofs) in appendix I. The Hilbert modular group $\Gamma_K = SL(2, \mathfrak{o}_K)$ (K totally real number field) has precisely h (=class number) cusp classes. The extended quotient $(\mathbf{H}^n)^*/\Gamma$ is compact. This result is also due to H. Maaß [48].

Let K be a field of complex numbers

$$\mathbf{Q} \subset K \subset \mathbf{C} \,.$$

We assume that K is an algebraic number field which means that the dimension of K as a vector space over $\mathbf{Q}$ is finite. This dimension is called the degree of K,

$$n := [K : \mathbf{Q}] := \dim_{\mathbf{Q}} K .$$

Such a number field admits n different imbeddings

$$K \longrightarrow \mathbf{C}; \quad a \longmapsto a^{(j)} \quad (1 \leq j \leq n)$$

into the field of complex numbers (An imbedding of K is an isomorphism of K onto a subfield of $\mathbf{C}$).

We assume that K is **totally real**, i.e. the image $K^{(j)}$ of each of the n imbeddings is contained in the field $\mathbf{R}$ of real numbers

$$K \overset{\sim}{\longrightarrow} K^{(j)} \subset \mathbf{R}$$
$$a \longmapsto a^{(j)} \quad (1 \leq j \leq n) .$$

We put the n imbeddings together into a single $\mathbf{Q}$-linear, obviously injective mapping

$$K \longrightarrow \mathbf{R}^n , \quad a \longmapsto (a^{(1)}, \ldots, a^{(n)}) .$$

For the sake of simplicity it is sometimes useful to identify a and the vector $(a^{(1)}, \ldots, a^{(n)})$.

If we attach to the matrix

$$M = \begin{pmatrix} a & b \\ c & d \end{pmatrix} \in GL(2, K)$$

the tuple

$$(M^{(1)}, \ldots, M^{(n)}), \quad M^{(j)} = \begin{pmatrix} a^{(j)} & b^{(j)} \\ c^{(j)} & d^{(j)} \end{pmatrix}, \quad j = 1, \ldots, n ,$$

we obtain an imbedding of groups

$$GL(2, K) \hookrightarrow GL(2, \mathbf{R})^n .$$

Also we occasionally identify $GL(2, K)$ with its image.

3.1 Definition. *The Hilbert modular group of the totally real number field K is*

$$\Gamma_K = SL(2, \mathfrak{o}) ,$$

where $\mathfrak{o}$ denotes the ring of (algebraic) integers of K.

We recall that the set of all algebraic integers contained in K forms a ring with field of fractions K. Moreover the image of $\mathbf{o}$ under the imbedding

$$K \longrightarrow \mathbf{R}^n$$
$$a \longmapsto (a^{(1)}, \ldots, a^{(n)})$$

is a **lattice** of $\mathbf{R}^n$. We have especially

$$\mathbf{o} \cong \mathbf{Z}^n$$

as additive groups. The discreteness of $\mathbf{o}$ in $\mathbf{R}^n$ implies the discreteness of $SL(2, \mathbf{o})$ in $SL(2, \mathbf{R})^n$. From 2.1 we obtain:

3.2 Remark. *The Hilbert modular group* $\Gamma_K = SL(2, \mathbf{o})$ *of a totally real number field K acts discontinuously on the product of n upper half-planes.*

We want to determine the cusps of the Hilbert modular group. The stabilizer of ∞ consists of all transformations of the form

$$z \longmapsto \varepsilon^2 z + a, \ a \in \mathbf{o}, \ \varepsilon \in \mathbf{o}^*$$

($\mathbf{o}^*$ is the multiplicative group of units, i.e. invertible elements of $\mathbf{o}$). From this we see:

a) The translation module of the Hilbert modular group is $\mathbf{o}$, hence isomorphic to $\mathbf{Z}^n$:

$$\mathbf{t} = \mathbf{o} \cong \mathbf{Z}^n .$$

b) The group of multipliers Λ is the group of squares of units

$$\Lambda = \{\varepsilon^2 \mid \varepsilon \in \mathbf{o}^*\} .$$

In the case of a totally real field the famous **Dirichlet unit theorem** states

$$\mathbf{o}^* \cong \mathbf{Z}^{n-1} \times \mathbf{Z}/2\mathbf{Z} .$$

From this it follows immediately that

$$\Lambda \cong \mathbf{Z}^{n-1} .$$

What we have shown is that ∞ is cusp of the Hilbert modular group.
Before we determine the other cusps, we mention a simple fact:
Let $\Gamma_0 \subset \Gamma$ be a subgroup of finite index. If Γ has cusp ∞, then Γ_0 also has cusp ∞ (and conversely). (This of course implies that Γ_0 and Γ have the same cusps.) The mentioned fact follows from two easy group theoretical lemmas.

Lemma 1. *Let G be a group and $G_0 \subset G$ a subgroup of finite index. Let $H \subset G$ be any subgroup of G. Then $H_0 = H \cap G_0$ has finite index in H (because the natural map of cosets*

$$H/H_0 \longrightarrow G/G_0$$
$$hH_0 \longmapsto hG_0$$

is injective).

Lemma 2. *Let $\mathbf{t} \subset \mathbf{Z}^n$ be a subgroup of finite index. Then*

$$\mathbf{t} \cong \mathbf{Z}^n.$$

Important examples of normal subgroups of finite index of the Hilbert modular group are the **principal congruence subgroups**: Let $\mathbf{a} \subset \mathbf{o}$ be an ideal (different from 0) in $\mathbf{o}$. The principal congruence subgroup of level $\mathbf{a}$ is the kernel of the natural homomorphism

$$SL(2, \mathbf{o}) \longrightarrow SL(2, \mathbf{o}/\mathbf{a}).$$

The group $SL(2, \mathbf{o}/\mathbf{a})$ is finite because $\mathbf{o}/\mathbf{a}$ is a finite ring. Therefore the kernel is a normal subgroup of finite index. We denote it by

$$\Gamma_K[\mathbf{a}] = \{M \in \Gamma_K \mid M \equiv E \bmod \mathbf{a}\}.$$

3.3 Remark. *Let be $A \in GL(2, K)$. There exists an ideal $\mathbf{a} \subset \mathbf{o}$ different from 0 such that*

$$\Gamma_K[\mathbf{a}] \subset A\Gamma_K A^{-1}.$$

Proof. There exists a natural number l such that the matrices lA and lA^{-1} have integral coefficients. The ideal $\mathbf{a} = (l^2)$ has the desired property. $\square$

Two subgroups Γ_1, Γ_2 of a given group are called **commensurable** if $\Gamma_1 \cap \Gamma_2$ has finite index in both the groups Γ_1, Γ_2. An immediate consequence of 3.3 is

3.3_1 Corollary. *If $\Gamma \subset SL(2, \mathbf{R})^n$ is any group commensurable with Hilbert's modular group, then the same is true of*

$$A\Gamma A^{-1}, \quad A \in GL(2, K).$$

3.4 Proposition. *The cusps of the Hilbert modular group are the elements of* $K \cup \{\infty\}$.

(We recall that an element $a \in K$ has to be identified with the vector $(a^{(1)}, \ldots, a^{(n)}) \in \mathbf{R}^n$.)

Proof. 1) Let be $a \in K$. The matrix

$$A = \begin{pmatrix} 0 & 1 \\ -1 & a \end{pmatrix} \in GL(2, K)$$

transforms a to ∞. By 3.2 ∞ is a cusp of $A\Gamma_K A^{-1}$ and therefore a is a cusp of Γ_K.
2) Let

$$\kappa = (\kappa_1, \ldots, \kappa_n) \in \overline{\mathbf{R}}^n$$

be a cusp of Γ_K. The stabilizer of κ contains a conjugate of a non-trivial translation and therefore an element

$$M = \begin{pmatrix} a & b \\ c & d \end{pmatrix} \neq \pm E, \quad a + d = \pm 2.$$

The fixed point equation

$$M^{(j)} \kappa_j = \kappa_j, \quad j = 1, \ldots, n$$

has only one solution, namely

$$\kappa = \frac{(d - a)}{2c} \in K. \qquad \qquad \square$$

Cusp Classes. We have to clarify when two cusps are equivalent under the action of the Hilbert modular group.

For this purpose we consider for each element $a \in K$ the ideal generated by a and 1

$$a \longmapsto (a, 1).$$

We extend this by sending ∞ to the principal ideal (1)

$$\infty \longmapsto (1).$$

We are not really interested in the constructed ideal but only in its ideal class. If we denote by $C(K)$ the (finite) set of ideal classes, the above construction gives a mapping

$$K \cup \{\infty\} \longrightarrow C(K)$$
$$a \longmapsto \begin{cases} \text{class of } (a,1) \text{ if } a \in K \\ \text{trivial class if } a = \infty. \end{cases}$$

3.5 Lemma. *The mapping*

$$K \cup \{\infty\} \longrightarrow C(K)$$

is surjective. Two cusps are equivalent with respect to the Hilbert modular group if and only if the corresponding ideal classes coincide.

This means that we have constructed a bijective mapping

$$K \cup \{\infty\}/\Gamma_K \overset{\sim}{\longrightarrow} C(K).$$

3.5₁ Corollary. *The Hilbert modular group has only finitely many cusp classes. Their number equals the class number of K.*

Proof. The proof of 3.5 will be an easy consequence of the following

3.5₂ Lemma. *Let*

$$\mathbf{a} = (c_1, d_1) = (c_2, d_2)$$

be an ideal of K. There exists a matrix

$$M \in \Gamma_K = SL(2, \mathbf{o})$$

with the property

$$(c_1, d_1) \cdot M = (c_2, d_2).$$

Proof. The (fractional) ideals of a number field form a multiplicative group! Therefore we have

$$1 \in \mathbf{o} = \mathbf{a} \cdot \mathbf{a}^{-1}$$

and can find elements

$$a_j, b_j \in \mathbf{a}^{-1}; \quad j = 1, 2$$

such that the matrices

$$M_j = \begin{pmatrix} a_j & b_j \\ c_j & d_j \end{pmatrix}; \quad j = 1, 2$$

have determinant 1. Obviously

$$M = M_1^{-1} M_2 = \begin{pmatrix} d_1 & -b_1 \\ -c_1 & a_1 \end{pmatrix} \begin{pmatrix} a_2 & b_2 \\ c_2 & d_2 \end{pmatrix}$$

has integral coefficients and therefore is contained in $SL(2, \mathfrak{o})$. We have

$$(c_1, d_1)M = (0, 1)M_2 = (c_2, d_2). \qquad\qquad \square$$

Proof of 3.5.
1) Let κ, κ' be equivalent cusps

$$\kappa' = M\kappa, \quad M \in SL(2, \mathfrak{o}).$$

We obtain

$$(\kappa, 1) = ((\kappa, 1) \cdot M') = (a\kappa + b, c\kappa + d) \sim (\kappa', 1)$$

where M' denotes the transposed matrix of M.

2) Assume that κ, κ' are cusps such that the corresponding ideals are equivalent:

$$(\kappa', 1) = \alpha \cdot (\kappa, 1).$$

By 3.5_2 we find a matrix $M \in SL(2, \mathfrak{o})$ such that

$$(\kappa', 1) = (\alpha\kappa, \alpha) \cdot M',$$

i.e. $\kappa' = M\kappa$. $\qquad\qquad\qquad\qquad\qquad\qquad\qquad\qquad \square$

The finiteness of the number of cusp classes is a hint, but of course no proof for

3.6 Theorem. *The space*

$$X_{\Gamma_K} = \mathsf{H}^n \cup K \cup \{\infty\} / \Gamma_K$$

is compact.

Proof. It is sufficient (because of 2.10) to construct finitely many cusps

$$\kappa_j = A_j^{-1}(\infty); \quad j = 1, \ldots, h$$

and a number $\delta > 0$ such that

$$\bigcup_{j=1}^{h} A_j^{-1}(U_\delta)$$

is a fundamental set of Γ_K. We arrange the proof in several lemmata (3.6_1–3.6_3).

3.6₁ Lemma. *There exists a constant $C = C(K)$ such that for each $x \in \mathbf{R}^n$ and for each $\varepsilon > 0$ we can find integers $c, d \in \mathbf{o}$, $c \neq 0$, which satisfy the inequalities*

$$||cx + d|| \leq \varepsilon, \quad ||c|| \leq \frac{C}{\varepsilon}.$$

Here we use the notation

$$||x|| = \max_{1 \leq j \leq n} |x_j|$$

and of course

$$cx = (c^{(1)}x_1, \ldots, c^{(n)}x_n).$$

Proof. Let $P \subset \mathbf{R}^n$ be a fundamental parallelotope of $\mathbf{o} \hookrightarrow \mathbf{R}^n$. We can find a finite number of ε-balls (with respect to the norm $||.||$) which cover P

$$P \subset U_1 \cup \ldots \cup U_N.$$

The number N of course depends on ε. But it is easy to see that one can achieve

$$N \leq \frac{N'}{\varepsilon^n}$$

where $N' = N'(K)$ depends only on K. We now choose the constant $C = C(K)$ such that the number of all

$$c \in \mathbf{o}, \quad ||c|| \leq \frac{C}{2\varepsilon}$$

is greater than $\frac{N'}{\varepsilon^n}$. This is possible because for any lattice $\mathbf{t} \subset \mathbf{R}^n$ there exists a constant $\delta > 0$ such that the number of points

$$a \in \mathbf{t}, \quad ||a|| \leq r$$

is greater or equal δr^n. We now choose for each c

$$c \in \mathbf{o}, \quad ||c|| \leq \frac{C}{2\varepsilon}$$

a $d \in \mathbf{o}$ such that

$$cx + d \in U_1 \cup \ldots \cup U_N.$$

At least one of the balls U_j has to contain two points $cx + d$ with different c. The difference of such two $cx + d$ is a solution of our inequalities. □

3.6₂ Lemma. *There is a constant* $\delta = \delta(K)$ *such that for each* $z \in H^n$ *there exists a pair*

$$(c,d) \in \mathfrak{o} \times \mathfrak{o}, \quad (c,d) \neq (0,0)$$

of integers which satisfies the inequality

$$Ny \geq \delta |N(cz+d)|^2 .$$

(Recall: $Nz = z_1 \cdot \ldots \cdot z_n$ for $z \in \mathbb{C}^n$)

Proof. We may replace z by $z \cdot \varepsilon^2, \varepsilon \in \mathfrak{o}^*$, and therefore assume that y is contained in a fundamental set with respect to the action

$$y \longmapsto y\varepsilon^2 , \quad \varepsilon \in \mathfrak{o}^* .$$

We therefore may assume that the inequalities

$$A^{-1} \leq \frac{y_j}{\sqrt[n]{Ny}} \leq A \quad (1 \leq j \leq n)$$

with a certain constant $A = A(K)$ are satisfied. We now take two integers as in Lemma 3.6₁ such that

$$\|cx + d\| \leq \sqrt[2n]{Ny} , \quad \|c\| \leq \frac{C}{\sqrt[2n]{Ny}} \quad (c,d \in \mathfrak{o} , \ c \neq 0) .$$

We obtain

$$|N(cz+d)|^2 = N[(cx+d)^2 + c^2 y^2]$$

$$\leq \prod_{j=1}^{n} \left[\sqrt[n]{Ny} + \frac{C^2 y_j^2}{\sqrt[n]{Ny}} \right]$$

$$\leq (1 + C^2 A^2)^n Ny,$$

and we can choose $\delta = (1 + C^2 A^2)^{-n}$. □

We want to deduce Theorem 3.6 from Lemma 3.6₂. It may be helpful for the reader first to treat the simple case of class number $h = 1$. In this case one may achieve in 3.6₂ that (c,d) is the second line of a modular matrix

$$M = \begin{pmatrix} a & b \\ c & d \end{pmatrix} \in SL(2,\mathfrak{o}) .$$

Then the inequality 3.6₂ is equivalent with

$$Ny/|cz + d|^2 = N(\text{Im } Mz) \geq \delta, \text{ i.e.} Mz \in U_\delta .$$

Therefore U_δ is a fundamental set.

To extend this proof to the case of arbitrary class numbers we have to prove a further lemma.

Take a set of representatives of the ideal classes

$$\mathbf{a}_j = (c_j, d_j); \quad j = 1, \ldots, h$$

and complete the pairs (c_j, d_j) to matrices of $SL(2, K)$

$$M_j = \begin{pmatrix} a_j & b_j \\ c_j & d_j \end{pmatrix} \in SL(2, K).$$

We may assume that all the $\mathbf{a}_j$ are integral ($\subset \mathbf{o}$).

3.6$_3$ Lemma. *There exists a constant $\varepsilon = \varepsilon(K)$ such that for each pair*

$$(c, d) \in \mathbf{o} \times \mathbf{o}, \quad (c, d) \neq (0, 0)$$

we can find
a) a matrix $M \in SL(2, \mathbf{o})$,
b) an index $j \in \{1, \ldots, h\}$ such that the following condition is satisfied

$$M_j M = \begin{pmatrix} * & * \\ \alpha c & \alpha d \end{pmatrix},$$

where

$$\|\alpha\| \leq \varepsilon^{-1}.$$

Proof. Let $\mathbf{a}_j$ be the ideal which lies in the same ideal class as (c, d): $\alpha \cdot (c, d) = \mathbf{a}_j$ for some $\alpha \in K$. We have

$$|N\alpha| \leq \mathcal{N}(\alpha c, \alpha d) = \mathcal{N}(\mathbf{a}_j).$$

We may replace α by $\alpha \varepsilon, \varepsilon \in \mathbf{o}^*$, and therefore assume that $\|\alpha\|$ is bounded from above by a certain constant which depends only on K. We now choose a matrix $M \in SL(2, \mathbf{o})$ such that

$$(c_j, d_j) \cdot M = (\alpha c, \alpha d) \quad \text{(see 3.5$_2$)}.$$

This means

$$\tilde{c} = \alpha c, \quad \tilde{d} = \alpha d$$

where $(\tilde{c}, \tilde{d})$ is the second line of $M_j M$.   □

Proof of Theorem 3.6. From 3.6$_2$ and 3.6$_3$ we obtain that for a given point $z \in \mathsf{H}^n$ there exists a matrix $M \in SL(2, \mathbf{o})$ and an index $j \in \{1, \ldots, h\}$ such that

$$Ny \geq \tilde{\delta}|N(cz + d)|^2,$$

where

$$M_j M = \begin{pmatrix} * & * \\ c & d \end{pmatrix}.$$

Here $\tilde{\delta}$ denotes some constant depending only on K. The above inequality is equivalent with

$$N(\operatorname{Im} M_j M z) \geq \tilde{\delta}.$$

The point Mz, which is Γ_K-equivalent with z, is therefore contained in $M_j^{-1}(U_\delta)$. The union of these domains is therefore a fundamental set.   □

An Application of 3.6. The Hilbert modular group Γ_K satisfies the condition of irreducibility 2.13. The same is true of each subgroup $\Gamma \subset SL(2,K)$ commensurable with Γ_K. From 3.6 we may therefore conclude

A subgroup $\Gamma \subset SL(2,K)$ which is commensurable with the Hilbert modular group Γ_K has only a finite number of classes of elliptic fixed points (as well as of cusps).

As an immediate consequence of this and of 2.14_1 we obtain

3.7 Remark. *The Hilbert modular group (more generally, any commensurable group) has only a finite number of conjugacy classes of elements of finite order.*

(When G is a group, then the conjugacy class of an element $a \in G$ is the set $\{xax^{-1} \mid x \in G\}$.)

In this context we mention

3.8 Remark. *The principal congruence subgroup*

$$\Gamma_K[l] = \ker(\Gamma_K \longrightarrow SL(2, \mathfrak{o}/(l)))$$

contains no element of finite order different from E if $l \geq 3$ $(l \in \mathsf{N})$.

Proof. Each finite group $G \neq \{e\}$ contains an element of prime order. If the statement of 3.8 is false, we can find a matrix

$$M \in \Gamma_K[l], \quad M \neq E, \quad M^p = E,$$

where p is a prime number. We write $M = E + B$ $(B \neq 0)$ and denote by $\mathfrak{b}$ the ideal generated by the coefficients of B. We have

$$\mathfrak{b} \subset (l) \quad (\text{because } M \in \Gamma_K[l]).$$

From $M^p = E$ we obtain

$$(*) \qquad \qquad \sum_{j=1}^{p} \binom{p}{j} B^j = 0.$$

The binomial coefficients being integral we obtain

$$pB \equiv 0 \bmod \mathbf{b}^2,$$

hence

$$p\mathbf{b} \subset \mathbf{b}^2$$

and therefore

$$p \in \mathbf{b} \subset (l).$$

The binomial coefficients $\binom{p}{j}, 1 \leq j \leq p-1$ are divisible by p. The coefficients of $\binom{p}{j}B^j$ are therefore contained in

$$p\mathbf{b}^2 \quad (\subset \mathbf{b}^3) \quad \text{if } j \geq 2.$$

From our assumption we obtain $p \geq 3$. Hence the coefficients of B^p are also contained in $\mathbf{b}^3$. From the equation $(*)$ we now obtain

$$pB \equiv 0 \bmod \mathbf{b}^3.$$

This implies

$$\mathbf{b} \subset p^{-1}\mathbf{b}^3$$

or equivalently

$$p \in \mathbf{b}^2 \subset (l^2).$$

But this is impossible since l^2 cannot divide p.   □

3.9 Corollary. $\Gamma[l]$ *acts freely on* $\mathbf{H}^n$ *for* $l \geq 3$, *i.e. there are no elliptic fixed points.*

§4 Automorphic Forms

Let

$$r = (r_1, \ldots, r_n) \in \mathbf{Z}^n$$

be a vector of rational integers. For

$$M \in SL(2, \mathbf{R})^n \quad \text{and} \quad z \in \mathbf{H}^n$$

we put

$$\mathcal{I}(M, z) = N(cz + d)^{2r} := \prod_{j=1}^{n} (c_j z_j + d_j)^{2r_j}.$$

This is a so-called factor of automorphy, i.e.

$$\mathcal{I}(MN, z) = \mathcal{I}(M, Nz) \cdot \mathcal{I}(N, z).$$

This condition makes it reasonable to consider functions

$$f : H^n \longrightarrow \mathbf{C}$$

with transformation law

$$f(Mz) = \mathcal{I}(M, z)f(z)$$

for all M which are contained in some given subgroup $\Gamma \subset SL(2, \mathbf{R})^n$.

The Fourier Expansion of f at a Cusp. We first assume that ∞ is a cusp of our (discrete) group Γ and denote by

$$\mathbf{t} = \{a \in \mathbf{R}^n \mid z \mapsto z + a \text{ lies in } \Gamma\}$$

the translation lattice of Γ, introduced in §2. From the translation formula

$$f(Mz) = \mathcal{I}(M, z)f(z)$$

we obtain

$$f(z + a) = f(z) \quad \text{for} \quad a \in \mathbf{t}.$$

We recall some basic facts about the Fourier expansion of holomorphic functions:

Let $\mathbf{t}$ be any lattice in $\mathbf{R}^n$. The **dual lattice** is defined as

$$\mathbf{t}^0 = \{a \in \mathbf{R}^n \mid S(ax) \in \mathbf{Z} \text{ for all } x \in \mathbf{t}\}. \quad (S = \text{trace})$$

Notice. Each lattice can be written in the form $\mathbf{t} = A\mathbf{Z}^n$, $A \in GL(n, \mathbf{R})$. An easy calculation shows $\mathbf{t}^0 = A'^{-1}\mathbf{Z}^n$.

Let $V \subset \mathbf{R}^n$ be any open (connected) domain. Then

$$D := \{z \in \mathbf{C}^n \mid y \in V\}$$

is the so-called tube domain corresponding to V.

4.1 Lemma. *Let*

$$f : D \longrightarrow \mathbf{C}$$

be any holomorphic function on the tube domain D over V. Assume

$$f(z + a) = f(z), \quad a \in \mathbf{t},$$

where $t \subset \mathbf{R}^n$ *is some lattice. Then f has a unique Fourier expansion*

$$f(z) = \sum_{g \in t^0} a_g e^{2\pi i S(gz)}.$$

(The summation is taken over the dual lattice). The series converges absolutely and uniformly on compact subsets. If $y \in V$ is an arbitrary point of V, the formula

$$a_g = \operatorname{vol}(P)^{-1} \int_P f(z) e^{-2\pi i S(gz)} \, dx$$

$$(z = x + iy, \quad dx = dx_1 \cdot \ldots \cdot dx_n)$$

holds. (vol(P) denotes the Euclidean volume of a fundamental parallelotope P of t*.)*

We now return to the special case where t is the translation lattice of a discrete subgroup of $SL(2, \mathbf{R})^n$ with cusp ∞.

4.2 Remark. *If t is the translation lattice of a discrete subgroup $\Gamma \subset SL(2, \mathbf{R})^n$ with cusp ∞, then each of the projections*

$$t \longrightarrow \mathbf{R}, \quad t^0 \longrightarrow \mathbf{R}$$

$$a \longmapsto a_j, \quad 1 \le j \le n$$

is injective. Especially

$$a \ne 0, \, a \ge 0 \implies a > 0.$$

$$(a \ge 0 \quad means: \quad a_j \ge 0 \quad for \, 1 \le j \le n)$$
$$(a > 0 \quad means: \quad a_j > 0 \quad for \, 1 \le j \le n)$$

Proof. See AI.19,1).   □

4.3 Definition. *Let $\Gamma \subset SL(2, \mathbf{R})^n$ be a discrete subgroup with cusp ∞ and $f : \mathbf{H}^n \to \mathbf{C}$ be a holomorphic function which is periodic with respect to the translation lattice of Γ. We call f* **regular** *at the cusp ∞ if the Fourier coefficients satisfy the condition*

$$a_g \ne 0 \implies g \ge 0.$$

We say that f **vanishes** *at cusp ∞ if it is regular and moreover $a_0 = 0$, equivalently:*

$$a_g \ne 0 \implies g > 0.$$

We want to generalize this to an arbitrary cusp. Assume that

$$f : \mathbf{H}^n \longrightarrow \mathbf{C}$$

is a holomorphic function with the transformation law

$$f(Mz) = N(cz + d)^{2r} f(z); \ (r = (r_1, \ldots, r_n) \in \mathbf{Z}^n)$$

for all M in our discrete subgroup Γ. Let $\kappa \in \overline{\mathbf{R}}^n$ be a cusp of Γ. We transform it to ∞:

$$A\kappa = \infty, \quad A \in SL(2, \mathbf{R})^n .$$

Notation:

$$\Gamma_A = A\Gamma A^{-1},$$
$$f_A(z) = \mathcal{I}(A^{-1}, z)^{-1} f(A^{-1}z),$$
$$\mathcal{I}(A, z) = N(cz + d)^{2r} .$$

A simple calculation shows

$$f_A(Mz) = \mathcal{I}(M, z) f_A(z) \quad \text{for} \quad M \in \Gamma_A .$$

The conjugate group Γ_A has cusp ∞ and we are tempted to say that f is regular at κ (vanishes at κ) if the same is true of f_A at ∞. But first we have to verify that this condition does not depend on the choice of A. Because two A's differ by a transformation which fixes ∞,

$$A\kappa = \infty, \ B\kappa = \infty \implies AB^{-1}(\infty) = \infty,$$

we only have to consider the special case

$$\kappa = \infty, \quad A\infty = \infty,$$

i.e.

$$A \sim \begin{pmatrix} 1 & b \\ 0 & 1 \end{pmatrix} \begin{pmatrix} a & 0 \\ 0 & a^{-1} \end{pmatrix} .$$

(The equivalence sign means that both sides define the same transformation, i.e. they have the same image in $(SL(2, \mathbf{R})/\{\pm E\})^n$.)

Let

$$f(z) = \sum_{g \in \mathfrak{t}^0} a_g e^{2\pi i S(gz)} , \quad f_A(z) = \sum_{g \in \mathfrak{t}_A^0} a_g^A e^{2\pi i S(gz)} ,$$

where

$$\mathfrak{t}_A = \mathfrak{t}(A\Gamma A^{-1}) = a^2 \mathfrak{t} \quad (\implies \mathfrak{t}_A^0 = a^{-2} \mathfrak{t}^0),$$

be the Fourier expansions of f and f_A. A simple calculation shows

$$f_A(z) = N(a)^{-2r} \sum_{g \in \mathfrak{t}^0} a_g e^{-2\pi i S(a^{-2}bg)} e^{2\pi i S(ga^{-2}z)} .$$

We obtain

$$a_g^A \;=\; N(a)^{-2r} e^{-2\pi i S(bg)} a_{ga^2} \quad (g \in \mathfrak{t}_A^0)\,.$$

4.4 Definition. *Let $\Gamma \subset SL(2,\mathbf{R})^n$ be a discrete subgroup with cusp κ and $f : \mathbf{H}^n \to \mathbf{C}$ be a holomorphic function with transformation law*

$$f(Mz) = \mathcal{I}(M,z)f(z) \quad \text{for } M \in \Gamma,$$
$$\mathcal{I}(M,z) = N(cz+d)^{2r} \quad (r \in \mathbf{Z}^n)\,.$$

*We say that f is **regular** at the cusp κ (**vanishes** at κ) if for some matrix (and then for each matrix)*

$$A \in SL(2,\mathbf{R})^n\,, \quad A\kappa = \infty,$$

the transformed function f_A is regular at ∞ (vanishes at ∞) in the sense of 4.3.

Notice. If f is regular at the cusp ∞ , we denote

$$f(\infty) := a_0$$

the value of f at ∞. It is not possible to define the value at an arbitrary cusp $\kappa = A^{-1}(\infty)$ because $f_A(\infty)$ depends on A. But if one changes A, the change in $f_A(\infty)$ is only a non-zero factor.

4.4₁ Remark. *(Same notation as in 4.4.) If f is regular at a certain cusp κ (vanishes at κ), then the same applies to each equivalent cusp.*

This follows immediately from the independence of the choice of A in 4.4. If for example $\kappa \sim \infty$ we may choose the transformation $A, A\kappa = \infty$ in Γ, which implies $f = f_A$.

4.5 Definition. *Let $\Gamma \subset SL(2,\mathbf{R})^n$ be a discrete subgroup such that the quotient $(\mathbf{H}^n)^*/\Gamma$ is compact. An **automorphic form** of weight $2r$, $r = (r_1,\dots,r_n) \in \mathbf{Z}^n$ with respect to Γ is a holomorphic function*

$$f : \mathbf{H}^n \longrightarrow \mathbf{C}$$

with the properties

a) $f(Mz) = N(cz+d)^{2r} f(z) \quad (M \in \Gamma),$

b) f *is regular at the cusps.*

*If f vanishes at all the cusps, we call f a **cusp form**.*

*If Γ is commensurable with the Hilbert modular group of a totally real number field, such an automorphic form is usually called a **Hilbert modular form**.*

It is of course sufficient to verify condition b) only for a set of representatives of all the cusps.

Notation:

$[\Gamma, 2r]$ = linear space of all automorphic forms of weight $2r$, $r = (r_1, \ldots, r_n) \in \mathbf{Z}^n$.

$[\Gamma, 2r]_0$ = subspace of all cusp forms.

4.5₁ Remark. *If $\Gamma_0 \subset \Gamma$ is a subgroup of finite index, then Γ_0 satisfies the same assumptions as Γ (formulation 4.5). The cusps of Γ_0 and Γ are the same and we have*

$$[\Gamma, 2r] \subset [\Gamma_0, 2r]$$
$$[\Gamma, 2r]_0 = [\Gamma, 2r] \cap [\Gamma_0, 2r]_0$$

with $r \in \mathbf{Z}^n$.

In the definition of an automorphic form we only considered even weights. This is sufficient for our purposes, but seems to be unnatural. What is really necessary for the definition of an automorphic form? Of course one first needs an automorphy factor, i.e. a mapping

$$\mathcal{I} : \Gamma \times \mathbf{H}^n \longrightarrow \mathbf{C} - \{0\}$$

which is holomorphic in the first variable and satisfies the cocycle condition

$$\mathcal{I}(MN, z) = \mathcal{I}(M, Nz) \cdot \mathcal{I}(N, z).$$

But in the definition of regularity at a cusp we need the definition of $\mathcal{I}(A, z)$ for a larger set of $A's$, because we have to transform an arbitrary cusp to infinity, which is usually not possible in Γ. We usually need a larger group G

$$\Gamma \subset G \subset SL(2, \mathbf{R})^n$$

to transform arbitrary cusps to ∞.

Examples: In the case of Hilbert's modular group Γ_K we can take $G = SL(2, K)$. We need an extension of the automorphy factor $\mathcal{I}$ to a mapping

$$\mathcal{I} : G \times \mathbf{H}^n \longrightarrow \mathbf{C} - \{0\}$$

which is holomorphic in the second variable. But it does not need to satisfy the precise cocycle condition. It is sufficient to have

$$\mathcal{I}(AB, z) = \mathrm{const}(A, B) \cdot \mathcal{I}(A, Bz) \cdot \mathcal{I}(B, z)$$

with some non-zero constant $(A, B \in G)$.

In order to get a Fourier expansion at each cusp the following condition should be satisfied: For each $A \in G$ there exists a sublattice of finite index

$$\mathfrak{l} \subset \mathfrak{t}_A$$

such that

$$\mathcal{I}(M, z) = 1$$

if

$$Mz = z + l, \quad l \in \mathfrak{l}, \quad M \in G.$$

Example: Let $r = (r_1, \ldots, r_n)$ be a vector of rationals (not necessarily integers). We define

$$N(cz + d)^r = e^{2\pi i \sum r_j \log(c_j z_j + d_j)}$$

by the main branch of the logarithm. It is sometimes possible to define a mapping

$$v : \Gamma \longrightarrow \mathbf{C} - \{0\}$$

for a given group Γ such that

$$\mathcal{I}(M, z) = N(cz + d)^r \cdot v(M)$$

is an automorphic factor (if r is integral one can take any character of Γ for v, for example $v \equiv 1$. For arbitrary rational r the conditions are more involved. Such a system $v(M)$ is called a multiplier system of weight r with respect to Γ. In the case of certain subgroups of Hilbert's modular group multiplier systems of weight $(1/2, \ldots, 1/2)$ are known. A general theory of these multiplier systems seems to be difficult.

In the paper "Automorphy Factors of Hilbert's Modular Group" (Proc. of the International Colloquium on Discrete Subgroups of Lie Groups and applications to Moduli, Bombay, January 1973) it is proved that each automorphy factor of a group commensurable with a Hilbert modular group is of the form

$$\mathcal{I}(M, z) = v(M) \frac{h(Mz)}{h(z)} N(cz + d)^r, \quad r \in \mathbf{Q}^n, \ | v(M) | = 1,$$

where h is a holomorphic function without zeroes. The denominators of r are bounded.

Modular Forms of Weight 0. Let $\Gamma \subset SL(2, \mathbf{R})^n$ be a discrete subgroup with cusp ∞. Let

$$f(z) = \sum_{g \in \mathfrak{t}^0} a_g e^{2\pi i S(gz)}$$

be a Fourier series which converges in some domain $U_C = \{z \in \mathbf{H}^n, \ Ny > C\}$. We assume V to be a cusp sector at ∞. In V we have an estimate

$$\delta^{-1} \sqrt[n]{Ny} \geq y_j \geq \delta \sqrt[n]{Ny} \quad (\delta \text{ some constant})$$

and therefore obtain

$$\lim_{Ny \to \infty, z \in V} f(z) = a_0 .$$

If, moreover, f is invariant under the whole stabilizer Γ_∞, we obtain

$$\lim_{Ny \to \infty} f(z) = a_0$$

without any further restriction on z. As a special case we obtain

4.6 Remark. *Let f be an automorphic form of weight $0 = (0, \ldots, 0)$ with respect to Γ. Then f is Γ-invariant and therefore defines a function on H^n/Γ which we denote again by f. This function extends continuously to $(H^n)^*/\Gamma$.*

Proof. The preceding remark shows that f extends continuously to ∞. For the other cusps one uses the technique of "transformation to ∞".   □

For the rest of this section we assume that the extended quotient $(H^n)^*/\Gamma$ is compact.

An important corollary of remark 4.6 is

4.7 Proposition. *Each automorphic form f of weight 0 is constant.*

Proof. It follows from 4.6 that f attains its maximum in $(H^n)^*/\Gamma$. If f is a cusp form, it has to be attained in H^n. From the maximum principle it follows that f has to be a constant (which is equal to zero). In the general case we denote the values of f at the cusp classes by $b_1, \ldots, b_h$. We notice that

$$\prod_{j=1}^{h} (f(z) - b_j)$$

is again an automorphic form of weight 0. It vanishes, because it is a cusp form. We obtain that f is one of the constants b_j.   □

Now we want to investigate the effect of the multipliers on the Fourier coefficients of an automorphic form

$$f(z) = \sum_{g \in t^0} a_g e^{2\pi i S(gz)}$$

of weight $(2r_1, \ldots, 2r_n)$. Let ε be a multiplier, i.e.

$$z \longmapsto \varepsilon z + b$$

is contained in Γ for some b. From the equation

$$f(\varepsilon z + b) = \varepsilon_1^{-r_1} \cdot \ldots \cdot \varepsilon_n^{-r_n} f(z) = N(\varepsilon^{-r}) f(z)$$

we obtain

$$a_{g\varepsilon} = a_g e^{2\pi i S(gb)} N \varepsilon^r ,$$

especially

$$| a_{g\varepsilon} | = | a_g | \varepsilon_1^{r_1} \cdot \ldots \cdot \varepsilon_n^{r_n} .$$

We give two applications. The first one is obvious:

4.8 Remark. *If f is an automorphic form, but not a cusp form, then*

$$r_1 = \ldots = r_n.$$

The second application is the so-called Götzky-Koecher principle which states that in the case $n \geq 2$ the regularity condition at the cusps can be omitted in the definition of an automorphic form:

Let be $\Gamma \subset SL(2,\mathbf{R})^n$ be any discrete subgroup with cusp ∞. As usually we denote by $\mathbf{t}$ the translation lattice and by Λ the group of multipliers.

4.9 Proposition. *Let $n \geq 2$ and*

$$f(z) = \sum_{g \in \mathbf{t}^0} a_g e^{2\pi i S(gz)}$$

be holomorphic and periodic (with respect to $\mathbf{t}$) on some domain $U_C = \{z \mid Ny > C\}$. We assume that there is an estimation

$$\mid a_g \mid \leq \|\varepsilon\|^{-A} \mid a_{g\varepsilon} \mid \quad for \quad \varepsilon \in \Lambda$$

with some constant A. Then

$$a_g \neq 0 \implies g \geq 0.$$

Corollary. *In the case $n \geq 2$ the regularity condition b) in the definition of an automorphic form (4.5) can be omitted.*

Proof. Let $g \in \mathbf{t}$ be a translation such that $g_1 < 0$. We choose a multiplier ε such that

$$\varepsilon_1 > 1, \varepsilon_2 < 1, \ldots, \varepsilon_n < 1$$

and obtain

$$-S(g\varepsilon^m) \geq \mid g_1 \mid \cdot \varepsilon_1^m + C \quad (m \in \mathbf{N})$$

where C is independent of m. From the absolute convergence of the Fourier series of f we obtain the convergence of the subseries

$$\sum_{m=1}^{\infty} \mid a_g \mid \varepsilon_1^{mA} e^{2\pi |g_1| |\varepsilon_1^m}.$$

This obviously implies $a_g = 0$. $\qquad\qquad\qquad\qquad\qquad\qquad\qquad\qquad$ $\square$

Some important results about the spaces $[\Gamma, r]$ are based on the following remark which is an immediate consequence of the formula

$$\operatorname{Im} Mz = \frac{\operatorname{Im} z}{|\, cz + d\,|^2}\,, \quad M \in SL(2, \mathbf{R})\,.$$

4.10 Remark. *Let f be an automorphic form of weight $2r$ with respect to Γ. The function*

$$g(z) = |\, f(z)\,| \cdot Ny^r$$

is Γ-invariant and therefore defines a continuous function

$$\mathbf{H}^n/\Gamma \longrightarrow \mathbf{R},$$

which we also denote by g. If f is a cusp form, then g extends to a continuous function on the compactification

$$(\mathbf{H}^n)^*/\Gamma\,.$$

The values at the cusp classes have to be defined as 0.

Corollary. *If f is a cusp form, g attains a maximum in $\mathbf{H}^n/\Gamma$.*

Proof. The only fact which remains to be shown is that

$$\lim_{Ny \to \infty} g(z) = 0\,,$$

if f is a cusp form and if ∞ is a cusp of Γ. As g is Γ_∞-invariant, it is sufficient to take the limit in a cusp sector. We claim that the series

$$\sum Ny^r \cdot |\, a_g\,|\, e^{-2\pi S(gy)}$$

converges uniformly in a cusp sector. The reason is that in a cusp sector we have an estimate

$$Ny^r e^{-2\pi S(gy)} \leq e^{-2\pi \delta S(g)}$$

with a certain positive bound δ and that

$$\sum |\, a_g\,|\, e^{-2\pi \delta S(g)}$$

converges (consider $f(i\delta, \ldots, i\delta)$). The uniform convergence allows us to take the limits term by term. A similar estimation shows

$$\lim (Ny)^r e^{-2\pi S(gy)} = 0,$$

where Ny tends to ∞ and iy varies in a cusp sector. $\square$

An important application of remark 4.10 is

4.11 Proposition. *Let* $\Gamma \subset SL(2,\mathbf{R})^n$ *be a discrete subgroup such that* $(\mathbf{H}^n)^*/\Gamma$ *is compact. We assume that* Γ *has cusps (i.e.* $\mathbf{H}^n/\Gamma$ *is not compact). Let* f *be any automorphic form of weight* $2r$

$$r \neq 0, \quad r_1 \cdot \ldots \cdot r_n = 0.$$

Then f *vanishes identically.*

Corollary. *Each holomorphic differential form on* $\mathbf{H}^n$ *which is invariant with respect to* Γ *and whose degree* p *satisfies*

$$0 < p < n$$

vanishes identically.

Proof. By assumption not all the r_j are equal. Therefore (4.8) f is a cusp form and

$$g(z) = |f(z)|\,(Ny)^r$$

attains its maximum in a certain point $z^{(0)} \in \mathbf{H}^n$. We may assume $r_1 = 0$. Then the function $f(z)(Ny)^r$ is holomorphic in z_1 and we obtain from the **maximum principle** that

$$z_1 \longmapsto f(z_1, z_2^{(0)}, \ldots, z_n^{(0)})$$

is constant. Let a be an element of the translation lattice $\mathbf{t}$. (We may assume that ∞ is a cusp of Γ.) We obtain

$$
\begin{aligned}
f(z_1^{(0)}, \ldots, z_n^{(0)}) &= f(z_1, z_2^{(0)}, \ldots, z_n^{(0)}) \\
&= f(z_1 + a_1, z_2^{(0)} + a_2, \ldots, z_n^{(0)} + a_n) \\
&= f(z_1, z_2^{(0)} + a_2, \ldots, z_n^{(0)} + a_n).
\end{aligned}
$$

We now have to make use of the fact that the image of the projection

$$\mathbf{t} \longrightarrow \mathbf{R}^{n-1}$$
$$a \longmapsto (a_2, \ldots, a_n)$$

is dense in $\mathbf{R}^{n-1}$ (AI.19). We may conclude that f only depends on the imaginary part of z. But f is a holomorphic function and hence constant. The constant has to be 0 since f is a cusp form. $\qquad\square$

The corollary of 4.11 will be basic in the determination of the cuspidal part of the cohomology of $\mathbf{H}^n/\Gamma$. This will be done in Chap. III where an introduction to differential forms is also given. A holomorphic differential form ω is an expression

$$\omega = \sum_{1 \le i_1 < \ldots < i_p \le n} f_{i_1,\ldots,i_p} \, dz_{i_1} \wedge \ldots \wedge dz_{i_p}$$

where the components $f_{i_1,\ldots,i_p}$ are holomorphic functions. The Γ-invariance of ω means

$$f_{i_1,\ldots,i_p}(Mz) \prod_{\nu=1}^{p} dM_{i_\nu}/dz = f_{i_1,\ldots,i_p}(z).$$

Because of

$$dM(z)/dz = (cz+d)^{-2}$$

this means that f is an automorphic form of weight $(r_1, \ldots, r_n)$ where

$$r_j = \begin{cases} 0 & \text{for } j \notin \{i_1,\ldots,i_p\} \\ 2 & \text{elsewhere}. \end{cases}$$

In the cases $1 \le p \le n-1$ we can apply 4.11.   □

It is natural to ask whether 4.11 is also true in the (usually simpler) case where H^n/Γ is compact. To clarify this, we define

4.12 Second Condition of Irreducibility. *The image of Γ under each of the n projections*

$$\pi_j : SL(2,\mathbf{R})^n \longrightarrow SL(2,\mathbf{R})^{n-1} \quad , \ 1 \le i \le n \, ,$$

(cancelling off one component) is dense in $SL(2,\mathbf{R})^{n-1}$.

Of course the Hilbert modular group satisfies this condition (because of AI.19,2)). But there are also groups Γ with compact quotient satisfying this condition. An argument similar to the one in the proof of 4.11 shows:

4.13 Remark. *The statement of 4.11 is also true if $\Gamma \subset SL(2,\mathbf{R})^n$ is a discrete subgroup with compact quotient H^n/Γ if it satisfies the second condition of irreducibility (4.12).*

In the next section we shall construct for each Γ a non-vanishing cusp form h of a certain weight $(r,\ldots,r), r > 0$. Assuming this for a moment we obtain

4.14 Lemma. *Let f be an automorphic form of a certain weight $(r_1,\ldots,r_n)$. Assume that the irreducibility condition 4.12 is satisfied if Γ has no cusps. If one of the components r_j is negative, f vanishes identically.*

Proof. We apply 4.7, 4.11 or 4.13 to the automorphic form $f^r \cdot h^{-r_j}$.   □

§5 Construction of Hilbert Modular Forms

I Construction of Cusp Forms. Let $D \subset \mathbf{C}^n$ be any (connected and open) domain and $\Gamma \subset \mathrm{Bihol}(D)$ a subgroup of the group of all biholomorphic mappings from D onto itself. We assume that Γ acts discontinuously (in the sense of 1.2). The group Γ is countable, because D can be written as the union of a countable set of compact subsets. Basic for the construction of cusp forms is the following

5.1 Lemma. *Let*

$$f : D \longrightarrow \mathbf{C}$$

be a holomorphic function such that

$$\int_D |f(z)|^2 \, dv$$

converges ($dv = dx_1 \cdot \ldots \cdot dx_n dy_1 \cdot \ldots \cdot dy_n$ denotes the usual Euclidean measure). The series

$$\sum_{\gamma \in \Gamma} |f(\gamma z)|^r |j(\gamma, z)|^r$$

then converges uniformly on compact subsets for $r \geq 2$.

In the following we denote by

$$j(\gamma, z) = \det(\partial \gamma_i / \partial z_j)$$

the Jacobian of an element $\gamma \in \mathrm{Bihol}(D)$. By the chain rule this is a factor of automorphy, i.e.

$$j(\gamma_1 \cdot \gamma_2, z) = j(\gamma_1, \gamma_2 z) \cdot j(\gamma_2, z).$$

Proof. We first notice that it is sufficient to prove the lemma in the case $r = 2$. We want to compare the series with an integral. For this we need

5.1₁ Lemma. *Let $K \subset D$ be a compact subset. There exists a constant C such that for each holomorphic function $f : D \to \mathbf{C}$ the inequality*

$$|f(a)|^2 \leq C \int_D |f(z)|^2 \, dv \quad \text{for } a \in K$$

is satisfied.

Proof. We choose a number $r > 0$ such that

$$U_r(a) := \{z \in \mathbf{C} \mid \|z - a\| < r\}$$

is contained in D for every $a \in K$. It is obviously sufficient to prove 5.1_1 for $U_r(a)$ instead of D and $\{a\}$ instead of K. We further may assume $a = 0$. We develop f into a power series

$$f(z) = \sum a_{j_1,\ldots,j_n} z_1^{j_1} \cdot \ldots \cdot z_n^{j_n} .$$

If one uses

$$\int_{x^2+y^2 \leq r^2} z^\mu \bar{z}^\nu \, dx dy = 0 \quad \text{for } \mu \neq \nu ,$$

one obtains

$$\int_{U_r(0)} |f(z)|^2 \, dv = \sum_{j_1,\ldots,j_n} \int_{U_r(0)} |a_{j_1\ldots j_n} z_1^{j_1} \cdot \ldots \cdot z_n^{j_n}|^2 dv$$

$$\geq (\pi r^2)^n \cdot |a_0|^2 = (\pi r^2)^n \cdot |f(0)|^2 . \qquad \square$$

To prove Lemma 5.1 it is obviously sufficient (because of 5.1_1) to construct for each $a \in D$ an open neighbourhood $U(a) \subset D$ such that

$$\sum_{\gamma \in \Gamma} \int_{U(a)} |f(\gamma z)|^2 |j(\gamma, z)|^2 dv$$

converges. We choose a neighbourhood $U(a)$ with the following two properties (compare 1.7)

a) $$\gamma(U(a)) \cap U(a) \neq \emptyset \implies \gamma \in \Gamma_a$$

b) $$\gamma \in \Gamma_a \implies \gamma(U(a)) = U(a) .$$

We now notice that $|j(\gamma, z)|^2$ is the real functional determinant of the transformation γ. (If $A : \mathbf{C}^n \to \mathbf{C}^n$ is a C-linear mapping with determinant $\det A$, then $|\det A|^2$ is the determinant of the underlying R-linear mapping $(\mathbf{C}^n \cong \mathbf{R}^{2n})$). By means of the **transformation formula for integrals** we obtain

$$\sum_{\gamma \in \Gamma} \int_{U(a)} |f(\gamma z)|^2 |j(\gamma, z)|^2 dv = \sum_{\gamma \in \Gamma} \int_{\gamma(U(a))} |f(z)|^2 dv$$

$$\leq (\#\Gamma_a) \cdot \int_D |f(z)|^2 dv < \infty . \qquad \square$$

We want to apply Lemma 5.1 to the Hilbert modular group and therefore need examples of square integrable holomorphic functions on $\mathbf{H}^n$.

5.2 Remark. *The integral*

$$\int_{\mathbf{H}} \frac{dv}{|z+a|^4} \quad (dv = dx dy)$$

converges for each $a \in \mathbf{H}$.

The proof can be done by direct calculation. Another way to see it is to make use of the fact that the area of the unit disc $\mathbf{E}$ is finite and to transform this area to an integral on $\mathbf{H}$ by means of the transformation $z \mapsto (z-a)(z-\bar{a})^{-1}$. An immediate consequence of 5.1, 5.2 is

5.3 Proposition. *Let $\Gamma \subset SL(2,\mathbf{R})^n$ be a discrete subgroup,*

$$\varphi : \mathbf{H}^n \longrightarrow \mathbf{C}$$

a bounded holomorphic function and

$$w \in \mathbf{H}^n, \quad r \geq 2 \, (r \in \mathbf{Z}).$$

The series

$$F(z) = F_\varphi^{(r)}(z) = \sum_{M \in \Gamma} \frac{\varphi(Mz)}{N(Mz - \overline{w})^{2r} N(cz+d)^{2r}}$$

converges absolutely and uniformly on compact subsets of $\mathbf{H}^n$. It therefore represents a holomorphic function on $\mathbf{H}^n$. This function has the transformation law

$$F(Mz) = N(cz+d)^{2r} F(z) \quad \text{for } M \in \Gamma.$$

Remark: There exist many bounded holomorphic functions on $\mathbf{H}^n$, namely: Consider a biholomorphic mapping

$$M_0 : \mathbf{H}^n \overset{\sim}{\longrightarrow} \mathbf{E}^n$$

and a polynomial $P(z_1, \ldots, z_n)$. Then

$$\varphi(z) = P(M_0 z)$$

is a bounded holomorphic function on $\mathbf{H}^n$.

For the proof of 5.3 it remains to show the transformation law. The easiest way to prove it is to use the formula

$$f|(M \cdot N) = (f|M)|N,$$

where

$$(f|M)(z) = j(M, z)^{-1} f(Mz).$$

Here $j(M, z)$ denotes any factor of automorphy. The series considered here are of the type

$$F = \sum f|M$$

and we obtain

$$F|N = \sum f|M|N = \sum f|MN = F.$$ □

Series of the type

$$\sum_{M \in \Gamma} f|M,$$

especially the series considered in 5.3, are called **Poincaré series**. They give us non-trivial examples of Hilbert modular forms.

We are now going to show that the Poincaré series (5.3) vanishes at the cusps. It is sufficient to treat the case of cusp ∞:

$$\lim_{Ny \to \infty} \sum_{M \in \Gamma} |N(Mz - \overline{w}) \cdot N(cz + d)|^{-2r} = 0.$$

The series under the limit has a remarkable symmetry. It remains unchanged if one interchanges z and w. This follows from the trivial

5.4 Remark. *Let be*

$$M = \begin{pmatrix} * & * \\ c & d \end{pmatrix} \in SL(2, \mathbf{R}), \quad \widetilde{M} = \begin{pmatrix} * & * \\ \tilde{c} & \tilde{d} \end{pmatrix} = M^{-1}.$$

We have

$$-\overline{(Mz - \overline{w})(cz + d)} = (\widetilde{M}w - \overline{z})(\tilde{c}w + \tilde{d}).$$

It is therefore sufficient to take the limit

$$N(\text{Im } w) \longrightarrow \infty \quad \text{for fixed } z$$

(instead of $Ny \to \infty$). If w varies in a certain cusp sector V (which is sufficient for our purpose), each term of the series tends to 0:

$$\lim_{\text{Im} w \to \infty, w \in V} N(Mz - \overline{w})^{-2} = 0.$$

We have to verify that formation of limit and summation can be interchanged. This is an immediate consequence of

5.5 Lemma. *For each $C > 0$ there exists a $\delta > 0$ such that*

$$|z + w|^2 \geq \delta|z + i|^2$$

for all $z \in H$ and

$$w \in H, \quad \operatorname{Im} w > C, \quad |\operatorname{Re} w| < C^{-1}.$$

The proof is an elementary exercise and can be left to the reader. The lemma shows that our series converges uniformly if w varies in a cusp sector, because it can be majorized up to a constant by its value at $w = i$.

5.6 Proposition. *Let $\Gamma \subset SL(2, \mathbf{R})^n$ be a discrete subgroup such that $(H^n)^*/\Gamma$ is compact. The Poincaré series $F(z)$ defined in 5.3 represents a cusp form of weight $2(r, \ldots, r)$. If $\varphi \neq 0$ and $w \in H^n$ are given, there exist infinitely many r for which $F_{\varphi^r}^{(r)}$ does not vanish.*

The only thing that remains to be proved is the statement about the non-vanishing of F. This is a consequence of an elementary lemma whose proof is left to the reader.

5.6₁ Lemma. *Assume that (a_n) is a sequence of complex numbers such that*

$$\sum a_n^r$$

converges absolutely for all $r \in N$ and such that the value is 0 for all but finitely many r. Then

$$a_n = 0 \quad \text{for all } n.$$

It should be mentioned that this type of argument can be refined to prove the following

Existence Theorems. *I) Let $a, b \in H^n$ be points which are inequivalent with respect to Γ. There exists a Poincaré series F (hence a cusp form) of suitable weight such that*

$$F(a) = 0, \quad F(b) = 1.$$

II) There exist $n + 1$ Poincaré series

$$F_0, \ldots, F_n$$

of a suitable common weight, which are algebraically independent.

This means that each polynomial $P(x_0, \ldots, x_n)$ with the property

$$P(F_0, \ldots, F_n) = 0$$

vanishes identically.

II Construction of Non-cusp Forms. We assume that ∞ is a cusp of the discrete subgroup $\Gamma \subset SL(2, \mathbf{R})^n$. We want to consider a series of the type

$$\sum N(cz + d)^{-2r}, \ r \in \mathbf{N}.$$

This series cannot converge if we extend the summation over the whole group Γ, because for infinitely many $M \in \Gamma$, namely all $M \in \Gamma_\infty$, we have

$$N(cz + d)^{-2r} = 1.$$

From the chain rule we even may conclude: The expression

$$\mathcal{I}(M, z) := N(cz + d)^{-2r}$$

depends only on the coset $\Gamma_\infty M$. We therefore define

$$E_{2r}(z) = \sum_{M \in \Gamma_\infty \backslash \Gamma} N(cz + d)^{-2r},$$

where the summation is taken over an arbitrary set of representatives of the cosets $\Gamma_\infty M$.

We call this the **Eisenstein series of weight** $2r$ with respect to the cusp ∞ of Γ.

5.7 Lemma. *Assume that ∞ is a cusp of Γ. The series*

$$\sum_{M \in \Gamma_\infty \backslash \Gamma} |N(cz + d)|^{-\sigma}$$

converges in $\mathbf{H}^n$ for all $\sigma > 2$. The convergence is uniform on each cusp sector of ∞, especially on compact subsets of $\mathbf{H}^n$.

Proof. We first notice: Assume that some point $z_0 \in \mathbf{H}$ and a certain constant $C > 0$ are given. There exists a constant $\varepsilon > 0$ such that

$$|cz + d| > \varepsilon |cz_0 + d|$$

for all real c, d and all z with

$$|x| \le C, \quad y \ge C^{-1}.$$

The proof is easy (compare 5.5) and can be left to the reader.

Assume that the series in 5.7 converges at some point $z_0 \in \mathbf{H}^n$. The preceding remark then shows that the series will converge uniformly on each cusp sector (§2) at ∞.

We now choose any open subset $U \subset \mathbf{H}^n$ with compact closure in $\mathbf{H}^n$ such that

$$M(U) \cap U \,=\, \emptyset$$

for $M \in \Gamma$, unequal to the identity. There exists a constant $C > 0$ such that

$$N(\operatorname{Im} M(z)) \,\leq\, C \quad \text{for } z \in U \,,\; M \in \Gamma \,.$$

We now choose a set $\mathcal{M}$ of representatives of the cosets $\Gamma_\infty M$. As we may apply

$$M \,\longmapsto\, NM \,, \quad N \in \Gamma_\infty \,,$$

we may assume that

$$\bigcup_{M \in \mathcal{M}} M(U)$$

is contained in a domain B of the following type

$$B \,=\, \{ z \in \mathbf{H}^n \mid Ny \leq C \,,\; \|x\| \leq A \,,\; A^{-1} \leq \|y\|/Ny \leq A \} \,,$$

where A is a suitable positive number.

Basic for our proof of convergence is: The integral

$$\int_B (Ny)^{\sigma/2} \frac{dv}{(Ny)^2} \,, \quad \sigma > 2 \,,$$

converges. This follows from the fact that the integral

$$\int_0^1 y^\alpha dy \,, \quad \alpha > -1 \,,$$

converges. We may replace B by the smaller domain

$$\bigcup_{M \in \mathcal{M}} M(U)$$

and obtain that the series

$$\sum_{M \in \mathcal{M}} \int_{M(U)} (Ny)^{\sigma/2} \frac{dv}{(Ny)^2}$$

converges. We now change the variables in each of the integrals

$$z \,=\, Mw \,.$$

The formulae

$$dM/dw \,=\, (cw + d)^{-2}$$

and

$$\text{Im } z = \frac{\text{Im } w}{\mid cw + d \mid^2}$$

show that the volume element $dv/(Ny)^2$ is invariant under $SL(2,\mathbf{R})^n$ (see also II,1.1) and that the above series equals

$$\sum_{M\in\mathcal{M}} \int_U \frac{(Ny)^{\sigma/2}}{\mid N(cz+d)\mid^\sigma} \frac{dv}{(Ny)^2} .$$

The functions Ny, Ny^{-1} are bounded on the compact set $\overline{U} \subset \mathbf{H}^n$. We therefore obtain that

$$\sum_{M\in\mathcal{M}} \int_U \mid N(cz+d)\mid^{-\sigma} dv$$

converges and this implies that

$$\sum_{M\in\mathcal{M}} \mid N(cz+d)\mid^{-\sigma} \quad (\sigma > 2)$$

converges on U.   □

5.8 Proposition. *Let $\Gamma \subset SL(2,\mathbf{R})^n$ be a discrete subgroup with cusp ∞. The Eisenstein series*

$$E_{2r}(z) = \sum_{M\in\Gamma_\infty\backslash\Gamma} N(cz+d)^{-2r}$$

converges for $r \geq 2$ ($r \in \mathbf{Z}$) and represents a holomorphic function on $\mathbf{H}^n$ with the transformation property

$$E_{2r}(Mz) = N(cz+d)^{2r} E_{2r}(z) .$$

The value at ∞ is 1

$$E_{2r}(\infty) = 1 ,$$

but E_{2r} vanishes at all cusps which are not Γ-equivalent to ∞.

Proof. The first part follows from 5.7.

The value at ∞:
We have to show

$$\lim_{Ny\to\infty} E_{2r}(z) = 1 .$$

The limit can be taken within a cusp sector of ∞. Because of 5.7 the limit can be computed term by term. We have

$$\lim_{Ny\to\infty} N(cz+d)^{-2r} = \begin{cases} 0 & \text{if } c \neq 0 \\ 1 & \text{if } c = 0 . \end{cases}$$

The second assertion is true since d^2 is a multiplier if $c = 0$ (in this case we have $M \in \Gamma_\infty$).

The behaviour at a cusp $\kappa = A^{-1}\infty$ which is not Γ-equivalent with ∞: We have to show that

$$(E_{2r}|A^{-1})(z) := N(cz + d)^{-2r} E_{2r}(A^{-1}z),$$

$$A^{-1} = \begin{pmatrix} * & * \\ c & d \end{pmatrix}$$

vanishes at ∞. We may write

$$E_{2r} = \sum_{M \in \Gamma_\infty \backslash \Gamma} 1 \mid M$$

and therefore

$$E_{2r}|A^{-1} = \sum_{M \in \Gamma_\infty \backslash \Gamma} 1 \mid MA^{-1}$$

$$(E_{2r}|A^{-1})(z) = \sum_{M \in (A\Gamma A^{-1})_\infty \backslash (\Gamma A^{-1})} N(cz + d)^{-2r}.$$

The same argument as at the beginning of the proof of 5.7 shows that this series converges uniformly in a cusp sector at ∞. It is therefore sufficient to show

$$\lim_{Ny \to \infty} N(cz + d)^{-2r} = 0$$

(in a cusp sector) or equivalently

$$c \neq 0 \quad \text{for} \quad M \in \Gamma A^{-1}.$$

If there were some

$$M = NA^{-1} \in \Gamma A^{-1}$$

with $c = 0$, we would have

$$M\infty = \infty, \quad \text{hence} \quad N\kappa = NA^{-1}\infty = \infty,$$

i.e. $\kappa \sim \infty \bmod \Gamma$. $\qquad\qquad\qquad\square$

How can we define an Eisenstein series with respect to an arbitrary cusp κ of Γ? Such an Eisenstein series should vanish at all cusps which are not equivalent with κ, but it should not vanish at κ. A natural procedure is

1) to transform κ to ∞, $A\kappa = \infty$,

2) to consider the Eisenstein series with respect to the cusp ∞ of the conjugate group $A\Gamma A^{-1}$,

3) to transform back this Eisenstein series by means of A.

The result is

$$
E_{2r}^A = \left(\sum_{M \in (A\Gamma A^{-1})_\infty \backslash A\Gamma A^{-1}} 1 \mid M \right) \mid A = \sum_{M \in (A\Gamma A^{-1})_\infty \backslash A\Gamma A^{-1}} 1 \mid MA ,
$$

or, explicitly

$$
E_{2r}^A(z) = \sum_{M \in (A\Gamma A^{-1})_\infty \backslash A\Gamma} N(cz + d)^{-2r} .
$$

Notice. The group $(A\Gamma A^{-1})_\infty = A\Gamma_\kappa A^{-1}$ acts on the set $A\Gamma$ by multiplication from the left. Therefore we find that the set $A\Gamma$ is a disjoint union of cosets $(A\Gamma A^{-1})_\infty \cdot M$, $M \in A\Gamma$. The summation in $E_{2r}^A(z)$ is taken over a set of representatives M.

5.9 Remark. *Let κ be a cusp of Γ. The Eisenstein series*

$$
E_{2r}^A(z) = \sum_{M \in (A\Gamma A^{-1})_\infty \backslash A\Gamma} N(cz + d)^{-2r}
$$

(with $A\kappa = \infty$) depends only (up to a constant factor) on the Γ-equivalence class of κ.

5.9₁ Corollary. *Assume that there exist only finitely many cusp classes (this is the case if $(\mathbf{H}^n)^*/\Gamma$ is compact) and denote by*

$$
\kappa_j = A_j^{-1}(\infty) , \quad 1 \le j \le h,
$$

a set of representatives. The Eisenstein series

$$
E_{2r}^{(j)}(z) := E_{2r}^{A_j}(z) , \quad 1 \le j \le h,
$$

are linearly independent. They generate a vector space

$$
\mathcal{E} = \mathcal{E}(\Gamma) = \sum_{j=1}^{h} C E_{2r}^{(j)}
$$

which depends on Γ only.

We call $\mathcal{E}$ the **space of Eisenstein series of Γ**.

Proof. That E_{2r}^A depends (up to a constant factor) only on the Γ-equivalence class of κ is a consequence of the following two observations:

1) $$E_{2r}^A = E_{2r} \quad \text{if} \quad A \in \Gamma$$

(because in this case $A\Gamma = A\Gamma A^{-1} = \Gamma$).

2) Let ∞ be a cusp of Γ and assume

$$A\infty = \infty.$$

In this case we have

$$(A\Gamma A^{-1})_\infty = A\Gamma_\infty A^{-1}$$

and therefore

$$E_{2r}^A = \sum_{M \in A\Gamma_\infty A^{-1}\backslash A\Gamma} 1 \mid M$$

$$= \sum_{M \in \Gamma_\infty\backslash\Gamma} 1 \mid AM$$

$$= \text{const} \cdot E_{2r},$$

because

$$1 \mid A = \text{const} \quad \text{if} \quad A(\infty) = \infty. \qquad \square$$

5.10 Proposition. *Let $\Gamma \subset SL(2,\mathbf{R})^n$ be a discrete subgroup such that $(\mathsf{H}^n)^*/\Gamma$ is compact. For each automorphic form*

$$f \in [\Gamma, 2(r, \ldots, r)], \quad r \geq 2,$$

there exists a unique element E in the space of Eisenstein series such that $f - E$ is a cusp form, in other words

$$[\Gamma, 2(r, \ldots, r)] = [\Gamma, 2(r, \ldots, r)]_0 \oplus \mathcal{E}.$$

5.10₁ Corollary.

$$\dim[\Gamma, 2(r, \ldots, r)] = \dim[\Gamma, 2(r, \ldots, r)]_0 + h$$

($h = $ numbers of cusp classes).

Final Remark. What happens in the case $r = 1$? It can be shown that Eisenstein series of weight $(2, \ldots, 2)$ can be defined by means of the limit

$$E_2(z) = \lim_{\sigma \to 0+} \sum_{M \in \Gamma_\infty\backslash\Gamma} N(cz + d)^{-2} \cdot \mid N(cz + d) \mid^\sigma.$$

It is clear that this limit transforms like an automorphic form. But there is a great difference between the cases $n = 1$ and $n \geq 2$.

1) $n \geq 2$: In this case $E_2(z)$ is an automorphic form with the same properties as formulated above for $E_{2r}(z)$, $r > 1$. We especially have

$$\dim[\Gamma, (2, \ldots, 2)] = \dim[\Gamma, (2, \ldots, 2)]_0 + h .$$

We shall prove this later for congruence subgroups of the Hilbert modular group (Chap. III, §4).

2) $n = 1$: In this case $E_2(z)$ is usually not a holomorphic function (but a non-analytic automorphic form in the sense of Maaß). If the set of cusps is not empty in the one-variable case, the equation

$$\dim[\Gamma, 2] = \dim[\Gamma, 2]_0 + h - 1$$

holds.

§6 The Finiteness of Dimension of a Space of Automorphic Forms

The aim of this section is to give a short and elementary proof of the following

6.1 Theorem. *Let $\Gamma \subset SL(2, \mathbf{R})^n$ be a discrete subgroup such that the extended quotient $(\mathbf{H}^n)^*/\Gamma$ is compact. The dimension of the space $[\Gamma, 2r]$ of automorphic forms of a given weight r, $r \in \mathbf{Z}$, is finite.*

In the following proof we make the further

Assumption. $\mathbf{H}^n/\Gamma$ *is not compact, i.e. there exist cusps.*

The case of a compact quotient is easier and we make some comments at the end of this section how to modify the proof in this case. The proof will result from the comparison of two different norms on the space of cusp forms $[\Gamma, 2r]_0$.

Norm 1: We have shown that the function

$$g(z) = | f(z) | N y^r$$

is Γ-invariant and attains its maximum in $\mathbf{H}^n$. We hence may define a norm by

$$\|f\|_\infty := \max_{z \in \mathbf{H}^n} g(z) .$$

Norm 2: We choose a set of representatives

$$\kappa_1, \ldots, \kappa_h$$

of the cusp classes and transform them to infinity

$$A_j \kappa_j = \infty, \quad 1 \leq j \leq h, \; A_j \in SL(2, \mathbf{R})^n.$$

We consider the lattice

$$t_j \subset \mathbf{R}^n$$

of translations of the conjugate group

$$A_j \Gamma A_j^{-1}$$

and denote by

$$P_j \subset \mathbf{R}^n$$

a fundamental parallelotope of t_j. Let δ be a positive number.

Notation:

$$V_j(\delta) = \{A_j^{-1}(z) \mid x \in P_j, \; y_1 \geq \delta, \ldots, y_n \geq \delta\}.$$

Remark. 1) If $\delta > 0$ is small enough, the set

$$V(\delta) = V_1(\delta) \cup \ldots \cup V_h(\delta)$$

is a fundamental set of Γ.

2) The integral

$$\int_{V_j(\delta)} d\omega, \quad d\omega = \frac{dv}{(Ny)^2}$$

$(dv = dx_1 \cdot \ldots \cdot dy_n = $ Euclidean measure$)$ converges.

Proof. 1) follows from 2.11 .

2) The volume element

$$d\omega = dv/(Ny)^2$$

is invariant with respect to transformations

$$z \longmapsto Mz, \quad M \in SL(2, \mathbf{R})^n.$$

We hence have to prove the existence of

$$\int_{\substack{x \in \text{ compact set} \\ y_j \geq \delta \text{ for } 1 \leq j \leq n}} d\omega \quad,$$

and this is a consequence of the convergence of

$$\int_\delta^\infty y^{-2} dy \quad (\delta > 0).$$

$\square$

Let

$$f, g \in [\Gamma, 2r]_0$$

be two cusp forms of the same weight. The function

$$\varphi(z) = f(z)\overline{g(z)} \cdot Ny^{2r}$$

is Γ-invariant and bounded. We hence may define for each $\delta > 0$ the Hermitean inner product

$$< f, g >_\delta = \int_{V(\delta)} \varphi(z) d\omega \,.$$

We obtain a family of norms

$$\|f\|_{2,\delta} = +\sqrt{< f, f >_\delta} \,.$$

(If δ is small enough in the sense that $V(\delta)$ is a fundamental set, all the norms $\|f\|_\delta$ are equivalent. They are in fact equivalent with the norm deduced from the so-called Petersson inner product:

$$< f, g > = \int_{\mathbf{H}^n/\Gamma} \varphi(z) d\omega$$

which will play a basic role in Chap. II (see 1.1). The equivalence of all these norms is a consequence of the **finiteness property** of $V(\delta)$, i.e. the set of all

$$M \in \Gamma, \quad M(V(\delta)) \cap V(\delta) \neq \emptyset$$

is finite. We do not need this and omit a proof.)

We now come to the announced comparison of norms.

6.2 Lemma. *If δ is small enough there exists a constant $A = A(\delta, \Gamma, r)$ such that*

$$\|f\|_\infty \leq A\|f\|_{2,\delta}$$

for all

$$f \in [\Gamma, 2r]_0 \quad (r \in \mathbf{N}^n \text{ fixed}).$$

We assume for a moment that the lemma has been proved and show

Lemma 6.2 $\Longrightarrow$ Theorem 6.1: Let

$$f_1, \ldots, f_m \in [\Gamma, 2r]_0$$

be a system of orthonormal vectors with respect to $< ., . >_\delta$, i.e.

$$< f_i, f_k >_\delta = \delta_{ik} \,.$$

For an arbitrary

$$f = \sum_{j=1}^{m} C_j f_j, \quad C_j \in \mathbf{C},$$

we obtain from the lemma

$$\left| \sum C_j f_j(z) N y^r \right| \leq A \sqrt{\sum |C_j|^2}.$$

If we specialize

$$C_j = \overline{f_j(z)},$$

we obtain

$$\sum |f_j(z)|^2 N y^r \leq A \sqrt{\sum |f_j(z)|^2}$$

or

$$\sum |f_j(z)|^2 N y^{2r} \leq A^2.$$

Integrating along $V(\delta)$ with respect to the measure $d\omega$ we obtain

$$m \leq A^2 \cdot \int_{V(\delta)} d\omega < \infty.$$

Proof of Lemma 6.2.

We choose $\delta > 0$ small enough such that $V(2\delta)$ is still a fundamental set. The function

$$h(z) = |f(z)|(Ny)^r$$

is Γ-invariant. It is hence sufficient to prove

$$h(z) \leq A(\delta) \cdot \sqrt{\int_{V(\delta)} h(\zeta)^2 d\omega}$$

for all $z \in V(2\delta)$. We prove a little more, namely that for each $j \in \{1, \ldots, h\}$

$$h(z) \leq A \sqrt{\int_{V_j(\delta)} h(\zeta)^2 d\omega}$$

for all $z \in V_j(2\delta)$.

It is of course sufficient to consider the case of cusp infinity, i.e. we may replace $V_j(\delta)$ by

$$V_\infty(\delta) = \{z \mid x \in P, y_1 \geq \delta, \ldots, y_n \geq \delta\},$$

where P is a fundamental parallelotope of the translation lattice of Γ_∞. We now compute the integral under the root sign by means of the Fourier expansion of f:

$$f(z) = \sum_{g>0} a_g e^{2\pi i S(gz)} \, .$$

A simple calculation gives

$$\int_{V_\infty(\delta)} h(z)^2 d\omega = \sum |a_g|^2 \int_{V_\infty(\delta)} e^{-4\pi S(gy)}(Ny)^{2r} d\omega \, .$$

By means of the inequality

$$\int_\delta^\infty e^{-ay} y^r dy \geq r! \cdot e^{-a\delta} a^{-(r+1)}$$

(integration by parts!) we obtain the estimation

$$(*) \quad \int_{V_\infty(\delta)} h(z)^2 d\omega \geq \mathrm{vol}(P) \cdot (2r-2)! \cdot \sum |a_g|^2 \, e^{-4\pi\delta S(g)} \cdot N(4\pi g)^{-2r+1} \, .$$

On the other hand we obtain

$$h(z) \leq \sum |a_g| \, e^{-2\pi S(gy)}(Ny)^r \, .$$

The Cauchy-Schwartz inequality gives us

$$h(z) \leq \left[\sum (|a_g| \, e^{-3/2 \, \pi S(gy)})^2\right]^{1/2} \cdot \left[\sum (e^{-\pi/2 \, S(gy)}(Ny)^r)^2\right]^{1/2} \, .$$

We now assume

$$z \in V_\infty(2\delta) \quad (\text{not only } \in V_\infty(\delta))$$

and obtain

$$(**) \qquad\qquad h(z) \leq B \cdot \sqrt{\sum |a_g|^2 \, e^{-6\pi\delta S(g)}}$$

with a certain constant B. If we apply 2.9_1 to the set

$$\Delta = \begin{pmatrix} a & b \\ c & d \end{pmatrix} (a,b,c,d \in \mathfrak{t}^0) \, ,$$

we obtain that $|Ng|$ has a positive lower bound. This implies

$$e^{-6\pi\delta S(g)} \leq C e^{-4\pi\delta S(g)} N(4\pi g)^{-2r+1}$$

(C a suitable constant). Comparing $(*)$ and $(**)$ we obtain the desired inequality. $\square$

Final Remark. (case of compact quotient)

If $\mathbf{H}^n/\Gamma$ is compact, we can find a compact fundamental set $K \subset \mathbf{H}^n$. We furthermore can find a positive number such that the union of all discs

$$U(a;\varepsilon) := \left\{ z \; ; \; \left| \frac{z-a}{z-\overline{a}} \right| < \varepsilon \right\}$$

with $a \in K$ is still contained in a compact subset $\widetilde{K}$ of $\mathbf{H}^n$. We now replace in the proof above

$$V(2\delta) \quad \text{by} \quad K$$
$$V(\delta) \quad \text{by} \quad \widetilde{K}$$

and the Fourier expansion of f by the Taylor expansion with respect to the variable

$$w = \frac{z-a}{z-\overline{a}} \, .$$

A similar (even easier) calculation will give the inequality

$$\|f\|_\infty \leq A \cdot \int_K |f(z)|^2 \, N y^{2r} d\omega \, ,$$

analogous to 6.2.

Chapter II. Dimension Formulae

§1 The Selberg Trace Formula

We are going to express the dimension of the space of cusp forms of weight $2(r,\ldots,r), r > 1$, by a certain integral along a fundamental domain of the given group Γ. The function to be integrated is an infinite series derived from the Poincaré series considered in Chap. I, §5.

Basic for the trace formula is the so-called **kernel function** on H^n:

$$k(z,w) = N\left(\frac{z-\overline{w}}{2i}\right)^{-2}$$
$$= \prod_{j=1}^{n}\left(\frac{z_j - \overline{w}_j}{2i}\right)^{-2}.$$

1.1 Remark. *The kernel function has the transformation property*

$$k(Mz, Mw)j(M,z)\overline{j(M,w)} = k(z,w),$$

where

$$j(M,z) = N(cz+d)^{-2}.$$

The measure

$$dw_z = k(z,z)\,dz = \frac{dv_z}{(Ny)^2}$$

is invariant under the transformations

$$z \mapsto Mz, \quad M \in SL(2,\mathbf{R})^n.$$

$(dv_z := dx_1 \ldots dx_n dy_1 \ldots dy_n$ denotes the usual Euclidean measure.)
The proof of 1.1 is trivial.

In the following we denote by

$$\mathcal{L}_r = \mathcal{L}_r(\mathbf{H}^n), \quad r \in \mathbf{N},$$

the linear space of all holomorphic functions

$$f : \mathbf{H}^n \to \mathbf{C}$$

such that

$$|f(z)|(Ny)^r = |f(z)|k(z,z)^{-r/2}$$

is bounded. We notice that cusp forms of weight $2(r, \ldots, r)$ have this property.

1.2 Proposition. *Assume $r \geq 2$. Each function $f \in \mathcal{L}_r$ satisfies the integral equation*

$$f(z) = \left(\frac{2r-1}{4\pi} \right)^n \int_{\mathbf{H}^n} \left[\frac{k(z,w)}{k(w,w)} \right]^r f(w) \, d\omega_w \,.$$

(The integral is absolutely convergent.)

Proof. If $f \in \mathcal{L}_r(\mathbf{H}^n)$, then f is contained in $\mathcal{L}_r(\mathbf{H})$ as a function of each of its variables. It is obviously sufficient to prove the integral equation in the one-variable case. This will also be true of the proof of convergence. We therefore assume $n = 1$.

We transform the integral equation into the bounded model $\mathbf{E}$ (unit disc) by means of the transformation

$$\mathbf{H} \overset{\sim}{\longrightarrow} \mathbf{E}$$
$$w \mapsto \eta = (w - z)(w - \bar{z})^{-1}$$
$$(z \in \mathbf{H} \text{ is fixed}).$$

The inverse transformation is

$$\eta \mapsto w = (z - \bar{z}\eta)(1 - \eta)^{-1} \,.$$

Because of

$$dw/d\eta = 2iy(1 - \eta)^{-2} \quad (y = \text{Im } z)$$

the Euclidean volume elements transform like

$$dv_w \mapsto 4y^2 |1 - \eta|^{-4} dv_\eta \,.$$

On the other hand we have

$$\text{Im } w = y(1 - |\eta|^2)|1 - \eta|^{-2}$$

and hence the invariant volume element transforms like

$$d\omega_w \mapsto d\omega_\eta := \frac{4dv_\eta}{(1 - |\eta|^2)^2} .$$

We now introduce the function

$$g(\eta) = (1 - \eta)^{-2r} f(w(\eta)) \quad (\eta \in E) .$$

We have

$$g(0) = f(z) ,$$

and a straightforward calculation gives us the transformed integral equation:

$$g(0) = \frac{2r - 1}{4\pi} \int_E (1 - |\eta|^2)^{2r} g(\eta) \, d\omega_\eta$$

or

$$g(0) = \frac{2r - 1}{\pi} \int_E (1 - |\eta|^2)^{2r-2} g(\eta) \, dv_\eta$$

$(dv_\eta = $ Euclidean volume element$)$.

The function f is contained in $\mathcal{L}_r$ if the function

$$g(\eta)(1 - |\eta|^2)^r$$

is bounded. The integral converges if

$$\int_E (1 - |\eta|^2)^{r-2} dv_\eta < \infty .$$

If we introduce polar coordinates

$$\eta = \rho e^{i\varphi} ,$$
$$dv_\eta = \rho \, d\rho \, d\varphi ,$$

this turns out to be equivalent with

$$\int_0^1 (1 - \rho^2)^{r-2} \rho \, d\rho < \infty$$

This integral can be evaluated after the transformation $\sigma = \rho^2$ ($d\sigma = 2\rho \, d\rho$), and we obtain that it converges for $r > 1$. For the proof of the integral equation we make use of the **power series** of the holomorphic function g:

$$g(\eta) = \sum_{m=0}^\infty a_m \eta^m .$$

Once again we make use of polar coordinates and first integrate along the angle φ ($0 \leq \varphi \leq 2\pi$) for fixed $\rho < 1$. We may integrate term by term, because the power series converges uniformly on compact subsets of $\mathbf{E}$. But obviously

$$\int_0^{2\pi} \eta^m \, d\varphi = \begin{cases} 0 & \text{if } m > 0 \,, \\ 2\pi & \text{if } m = 0 \,. \end{cases}$$

For this reason the integral equation has only to be proved for constant functions g.

We have

$$\frac{2r-1}{\pi} \int_{\mathbf{E}} (1 - |\eta|^2)^{2r-2} dv_\eta = 2(2r-1) \int_0^1 (1 - \rho^2)^{2r-2} \rho \, d\rho$$

$$= (2r-1) \int_0^1 (1 - \sigma)^{2r-2} d\sigma = 1 \,. \qquad \square$$

We now consider a discrete subgroup $\Gamma \subset SL(2,\mathbf{R})^n$. Let l be the order of the kernel of the natural projection

$$\Gamma \to (SL(2,\mathbf{R})/\{\pm E\})^n \,.$$

(Two elements of Γ define the same transformation if their images coincide.) From the integral equation 1.2 we may deduce

$$f(z) = l^{-1} \left(\frac{2r-1}{4\pi}\right)^n \int_{\mathbf{H}^n/\Gamma} \sum_{M \in \Gamma} \left[\frac{k(z, Mw)}{k(Mw, Mw)}\right]^r f(Mw) \, d\omega_w \,.$$

We can replace $\mathbf{H}^n/\Gamma$ under the integral sign by a fundamental domain of Γ. The convergence of the inner series outside a neglectible set is a consequence of general facts about integration theory (AII.7). We shall obtain better information about the convergence without using this below.

We now assume that the extended space $X_\Gamma = (\mathbf{H}^n)^*/\Gamma$ (I, §2) is compact and that f is a cusp form of weight $2(r, \ldots, r)$ with respect to Γ. It follows from I.4.10 that f is contained in $\mathcal{L}_r$, and we may apply our integral equation to f. If we use the formulae

$$f(Mz) = j(M,z)^{-r} f(z)$$
$$(j(M,z) = N(cz+d)^{-2})$$

and

$$k(Mw, Mw) = |j(M,w)|^{-2} k(w,w) \,,$$

we obtain

$$f(z) = \left(\frac{2r-1}{4\pi}\right)^n \int_{\mathbf{H}^n/\Gamma} \frac{f(w)\overline{K(z,w)}}{k(w,w)^r} \, d\omega_w \,,$$

where
$$K(z,w) = K_{\Gamma,r}(z,w) = l^{-1} \sum_{M \in \Gamma} k(Mw,z)^r j(M,w)^r$$
$$= \frac{(2i)^{2rn}}{l} \sum_{M \in \Gamma} \frac{1}{N(Mw - \bar{z})^{2r} N(cw + d)^{2r}}$$

This type of series – a so-called Poincaré series – has been introduced in Chap. I, §5. From I.5.3, I.5.4 and I.5.6 we know

1.3 Proposition. *The function*
$$K(z,w) = K_{\Gamma,r}(z,w) = l^{-1} \sum_{M \in \Gamma} k(Mw,z)^r j(M,w)^r$$
$$\left(k(w,z) = N\left(\frac{w - \bar{z}}{2i}\right)^{-2}, \ j(M,w) = N(cw + d)^{-2} \right)$$

is (for fixed z) a cusp form of weight $2(r,\ldots,r)$ $(r \geq 2)$ as a function of w. It has the property
$$K(z,w) = \overline{K(w,z)}.$$

If f is a cusp form of weight $2(r,\ldots,r)$ we have
$$f(z) = \left(\frac{2r-1}{4\pi}\right)^n \int_{\mathbf{H}^n/\Gamma} \frac{f(w)\overline{K(z,w)}}{k(w,w)^r} d\omega_w.$$

We now introduce a Hermitean inner product in the space of cusp forms. We first notice

1.4 Remark. *The quotient space $\mathbf{H}^n/\Gamma$ has a finite volume with respect to the invariant measure $d\omega$.*
(We assume that $(\mathbf{H}^n)^*/\Gamma$ is compact!)

Proof. It is sufficient to construct a measurable fundamental set with finite volume. (AII.8, AII.10)

We use the fundamental set constructed in Chap. I, §2 (2.10) and have to show that each cusp sector has finite volume. We may restrict ourselves to the cusp ∞ and have therefore to show
$$\int_V \frac{dv}{(Ny)^2} < \infty$$

where V is a cusp sector at ∞. From the definition of a cusp sector it follows that there exists a constant $\delta > 0$ such that

a) $\|x\| \leq \delta^{-1}$

b) $y_j \geq \delta$ for $j = 1, \ldots, n$

for $z = x + iy \in V$.

The convergence of the integral now follows from the convergence of

$$\int_\delta^\infty \frac{dy}{y^2} < \infty \qquad (\delta > 0) \,. \qquad \qquad \square$$

We now consider two cusp forms f, g of a certain weight $2r, r = (r_1, \ldots, r_n)$. The function

$$f(z)\overline{g(z)}(Ny)^{2r}$$

is Γ-invariant and bounded (I4.10). We may conclude (1.4) that the integral

$$\langle f, g \rangle := \int_{\mathbf{H}^n/\Gamma} f(z)\overline{g(z)}(Ny)^{2r} d\omega_z$$

exists.

1.5 Remark. *The pairing $\langle f, g \rangle$ is a Hermitean inner product on the space of cusp forms $[\Gamma, 2r]_0$ of an arbitrary weight, i.e.*

    a) $\langle f, g \rangle$ *is $\mathbb{C}$-linear in f ,*

    b) $\langle f, g \rangle = \overline{\langle g, f \rangle}$,

    c) $\langle f, f \rangle > 0$ *for $f \neq 0$.*

Using this inner product we may rewrite the integral equation for cusp forms (1.3) in the form

$$\boxed{\; f(z) = \left(\frac{2r - 1}{4\pi}\right)^n \cdot \langle f, K(z, \bullet) \rangle \;}$$

We now choose an orthonormal basis $f_1, \ldots, f_m$ of the space $[\Gamma, 2(r, \ldots, r)]_0$. (We have already proved that this space is of finite dimension, I.6.1).

$$\langle f_i, f_j \rangle = \begin{cases} 1 & \text{for } i = j \\ 0 & \text{for } i \neq j \,. \end{cases}$$

We may express the kernel function

$$w \mapsto K(z, w)$$

by means of this basis:

$$K(z, w) = \sum_{j=1}^m a_j(z) f_j(w) \,.$$

The integral equation gives us

$$f_i(z) = \left(\frac{2r-1}{4\pi}\right)^n \langle f_i, K(z, \bullet)\rangle$$

$$= \left(\frac{2r-1}{4\pi}\right)^n \overline{a_i(z)},$$

i.e.

$$\left(\frac{2r-1}{4\pi}\right)^n K(z, w) = \sum_{j=1}^{m} \overline{f_j(z)} f_j(w).$$

We specialize this equation $(z = w)$, multiply it with $(Ny)^{2r}$ and integrate along $\mathbf{H}^n/\Gamma$. The result is the "trace formula"

1.6 Theorem. *Let $\Gamma \subset SL(2, \mathbf{R})^n$ be a discrete subgroup such that $(\mathbf{H}^n)^*/\Gamma$ is compact. Let $r \geq 2$ be a natural number. We have*

$$\dim[\Gamma, 2(r, \ldots, r)]_0 = \left(\frac{2r-1}{4\pi}\right)^n \int_{\mathbf{H}^n/\Gamma} \frac{K(z, z)}{k(z, z)^r}\, d\omega,$$

where

$$K(z, w) = l^{-1} \sum_{M \in \Gamma} k(Mw, z)^r j(M, w)^r,$$

$$k(w, z) = N\left(\frac{w - \bar{z}}{2i}\right)^{-2}, \quad j(M, w) = N(cw + d)^{-2},$$

$$d\omega = k(z, z)\, dv = \frac{dv}{(Ny)^2}$$

$(dv = \text{Euclidean volume element})$.

l is the order of the kernel of the natural projection $\Gamma \to (SL(2, \mathbf{R})/\{\pm E\})^n$.

Notice: Two elements $M, N \in \Gamma$ with the same image define the same term in the series of $K(z, w)$. The trace formula concerns rather the image of Γ (i.e. the underlying group of transformations) than Γ itself.

The Main Term of the Trace Formula. In the series defining $K(z, z)$ we extract all terms $M \in \Gamma$ which belong to the kernel of $\Gamma \to (SL(2, \mathbf{R})/\{\pm E\})^n$. We obtain

$$K(z, z) = k(z, z)^r + K'(z, z),$$

where

$$K'(z, z) = l^{-1} \sum_{\substack{M \in \Gamma \\ M \neq \text{ identity transformation}}} k(Mz, z)^r j(M, z)^r.$$

The trace formula can be written as

$$\dim[\Gamma, 2(r, \ldots, r)]_0 = \mathrm{vol}(\mathsf{H}^n/\Gamma)(2r - 1)^n + A(r),$$

where

a) $$\mathrm{vol}(\mathsf{H}^n/\Gamma) = (4\pi)^{-n} \int_{\mathsf{H}^n/\Gamma} d\omega$$

denotes the volume of H^n/Γ with respect to the invariant measure $(4\pi)^{-n} \, d\omega$.

b) $$A(r) = \left(\frac{2r-1}{4\pi}\right)^n \int_{\mathsf{H}^n/\Gamma} \frac{K'(z,z)}{k(z,z)^r} \, d\omega.$$

We shall see later that $A(r)$ plays the role of an **error term**, i.e.

$$\dim[\Gamma, 2(r, \ldots, r)]_0 \sim \mathrm{vol}(\mathsf{H}^n/\Gamma)(2r - 1)^n.$$

In this connection we would like to make some general remarks. The Selberg trace formula 1.6 can be generalized to an arbitrary arithmetic group Γ (instead of Hilbert's modular group) acting on a bounded symmetric domain $D \subset \mathbf{C}^n$ (instead of H^n). One has to replace $k(z, w)$ by the Bergman kernel function and $j(\gamma, z)$ ($\gamma \in \Gamma$) by the Jacobian. Instead of $(2r - 1)^n$ there occurs a certain polynomial $a(r)$ which is characteristic for the domain D. A cusp form of weight $r \in \mathbf{Z}$ is a holomorphic function $f : D \to \mathbf{C}$ with the transformation property

$$f(\gamma z) = j(\gamma, z)^{-r} f(z)$$

which vanishes at the cusps. There are two different methods to calculate the dimension of the space of cusp forms $[\Gamma, r]_0$, namely

    a)   the Selberg trace formula,

    b)   generalized Riemann-Roch theorems.

Both methods have been applied successfully to the case of a compact quotient D/Γ. It was Langlands, who did that in the first case and Hirzebruch in the second.

The case of a non-compact quotient D/Γ is much more involved. By means of the "Riemann-Roch method" Mumford proved

$$\dim[\Gamma, r]_0 = a(r) \cdot \mathrm{vol}(D/\Gamma) + S(r),$$
$$S(r) = O(r^{n-1}).$$

One may expect that this result admits the following improvement. Let X_Γ be the Baily-Borel compactification of D/Γ [4] (which generalizes our compactification in case of Hilbert's modular group). Let S be the locus of all elliptic fixed points and all boundary points. We expect

$$S(r) = O(r^d)$$

where d is the maximal dimension of an irreducible component of S.

§2 The Dimension Formula in the Cocompact Case

In this section we assume that $\Gamma \subset SL(2,\mathbf{R})^n$ is a discrete subgroup with compact quotient $\mathbf{H}^n/\Gamma$. We also assume that Γ satisfies the irreducibility condition I.2.13: Each of the n projections

$$\Gamma \to SL(2,\mathbf{R})$$

is injective. The trace formula 1.6 has the form

$$\dim[\Gamma, 2(r,\ldots,r)] = \left(\frac{2r-1}{4\pi}\right)^n l^{-1} \int_{\mathbf{H}^n/\Gamma} \left(\sum_{M\in\Gamma} k(M,z)\right) d\omega,$$

where

$$k(M,z) := \left[\frac{k(Mz,z)}{k(z,z)}\right]^r j(M,z)^r$$

and

$$k(z,w) = N\left(\frac{z-\overline{w}}{2i}\right)^{-2},$$

$$j(M,z) = N(cz+d)^{-2},$$

$l = $ order of the kernel of the image of Γ in $(SL(2,\mathbf{R})/\{\pm E\})^n$.

The series

$$\sum_{M\in\Gamma} |k(M,z)|$$

converges uniformly on compact subsets. We now write Γ as a disjoint union of conjugacy classes

$$[M_0] := \{M\,M_0\,M^{-1},\ M\in\Gamma\}.$$

We obtain

$$\sum_{M\in\Gamma} k(M,z) = \sum_{M_0}\left(\sum_{M\in[M_0]} k(M,z)\right),$$

where M_0 runs through a complete system of representatives of all conjugacy classes. We replace the domain of integration $\mathbf{H}^n/\Gamma$ by a precise (measurable) fundamental domain F. We may assume that F is contained in a compact subset of $\mathbf{H}^n$.

The uniform convergence on compact sets allows us to interchange summation and integration in the following way:

$$\dim[\Gamma, 2(r,\ldots,r)] = \left(\frac{2r-1}{4\pi}\right)^n l^{-1} \sum_{M_0} \int_F \sum_{M\in[M_0]} k(M,z)\,d\omega.$$

We call the occuring integral the contribution of the conjugacy class $[M_0]$ to the trace formula. We now simplify this contribution. We have

$$M \, M_0 \, M^{-1} = N \, M_0 \, N^{-1}$$

if $N^{-1} M$ is contained in the centralizer of M_0:

$$\Gamma_{M_0} = \{ M \in \Gamma \mid M \, M_0 = M_0 \, M \} \, ,$$

or equivalently if

$$M \, \Gamma_{M_0} = N \, \Gamma_{M_0} \, .$$

Whence we obtain

$$\sum_{M \in [M_0]} k(M, z) = \sum_{M \in \Gamma / \Gamma_{M_0}} k(M \, M_0 \, M^{-1}, z) \, ,$$

where M runs through a complete set of representatives of the cosets

$$M \, \Gamma_{M_0}, \ M \in \Gamma \, .$$

2.1 Lemma. *The function $k(M, z)$ satisfies the functional equation*

$$k(M \, M_0 \, M^{-1}, z) = k(M_0, M^{-1} z) \, .$$

We especially have that $z \mapsto k(M_0, z)$ is invariant under the centralizer Γ_{M_0}.

For the contribution of the conjugacy class $[M_0]$ we obtain the expression

$$\int_{\mathbf{H}^n / \Gamma_{M_0}} k(M_0, z) \, d\omega \, .$$

The advantage now is that a fundamental domain of Γ_{M_0} can be determined. For this purpose we determine the centralizer of an arbitrary element $M_0 \in SL(2, \mathbf{R})$, $M_0 \neq \pm E$,

$$Z(M_0) := \{ M \in SL(2, \mathbf{R}) \mid M \, M_0 = M_0 \, M \} \, .$$

Because of

$$Z(M \, M_0 \, M^{-1}) = M \, Z(M_0) \, M^{-1}$$

it is sufficient to restrict to a suitable system of representatives of the conjugacy classes.

2.2 Lemma. *An arbitrary element*

$$M_0 \in SL(2, \mathbf{R}), \ M_0 \neq \pm E$$

is conjugate in $SL(2,\mathbf{R})$ with

    1)   a translation matrix

$$\pm \begin{pmatrix} 1 & a \\ 0 & 1 \end{pmatrix} \qquad (\text{if } M_0 \text{ is parabolic}, i.e. \ |\sigma(M_0)| = 2),$$

    2)   a transvection matrix

$$\begin{pmatrix} \varepsilon & 0 \\ 0 & \varepsilon^{-1} \end{pmatrix} \qquad (\text{if } M_0 \text{ is hyperbolic}, i.e. \ |\sigma(M_0)| > 2),$$

    3)   an orthogonal matrix

$$\begin{pmatrix} \cos\varphi & \sin\varphi \\ -\sin\varphi & \cos\varphi \end{pmatrix} \qquad (\text{if } M_0 \text{ is elliptic}, i.e. \ |\sigma(M_0)| < 2).$$

The centralizers are

$$Z\begin{pmatrix} 1 & a \\ 0 & 1 \end{pmatrix} = \left\{ \pm \begin{pmatrix} 1 & b \\ 0 & 1 \end{pmatrix} \mid b \in \mathbf{R} \right\},$$

$$Z\begin{pmatrix} \varepsilon & 0 \\ 0 & \varepsilon^{-1} \end{pmatrix} = \left\{ \pm \begin{pmatrix} a & 0 \\ 0 & a^{-1} \end{pmatrix} \mid a \neq 0 \right\},$$

$$Z\begin{pmatrix} \cos\varphi & \sin\varphi \\ -\sin\varphi & \cos\varphi \end{pmatrix} = SO(2,\mathbf{R}).$$

Proof.
1) A parabolic transformation has precisely one fixed point in $\overline{\mathbf{R}} = \mathbf{R} \cup \{\infty\}$ and this can be transformed to ∞.
2) A hyperbolic transformation has two fixed points in $\overline{\mathbf{R}}$ which can be transformed simultaneously to 0 and ∞.
3) An elliptic transformation has a fixed point in $\mathbf{H}$. We can transform this fixed point to i.

The computation of the centralizers is trivial.   □

    We now investigate a fundamental domain of Γ_{M_0} where $\Gamma \subset SL(2,\mathbf{R})^n$ is our discrete subgroup. For our purposes it is always sufficient to replace Γ_{M_0} by a subgroup of finite index

$$\Gamma'_{M_0} \subset \Gamma_{M_0}$$

because of

$$\int_{\mathbf{H}^n/\Gamma'_{M_0}} k(M_0, z)\, d\omega = [\Gamma_{M_0} : \Gamma'_{M_0}] \int_{\mathbf{H}^n/\Gamma_{M_0}} k(M_0, z)\, d\omega \,.$$

We also notice:

If we replace M_0 by $M M_0 M^{-1}$, $M \in SL(2,\mathbf{R})^n$, and Γ by the group $M \Gamma M^{-1}$, then the integral will not change.

After this preparation we describe the fundamental domain. We may assume

$$M_0 = \Big(\underbrace{M_0^{(1)}, \ldots, M_0^{(k)}}_{\text{hyperbolic}}, \underbrace{M_0^{(k+1)}, \ldots, M_0^{(l)}}_{\text{parabolic}}, \underbrace{M_0^{(l+1)}, \ldots, M_0^{(n)}}_{\text{elliptic}} \Big)$$

and (after conjugation)

$$M_0^{(i)} = \begin{pmatrix} \varepsilon_i & 0 \\ 0 & \varepsilon_i^{-1} \end{pmatrix}; \quad 1 \le i \le k,$$

$$M_0^{(i)} = \begin{pmatrix} 1 & a_i \\ 0 & 1 \end{pmatrix}; \quad k < i \le l,$$

$$M_0^{(i)} \in SO(2,\mathbf{R}); \quad l < i \le n.$$

We now define a certain subgroup

$$\Gamma'_{M_0} \subset \Gamma_{M_0}$$

of finite index.

Case 1: $k + l = 0$ (all the components are elliptic).

In this case Γ_{M_0} is a discrete subgroup of the compact group $SO(2,\mathbf{R})^n$, hence finite. We may take

$$\Gamma'_{M_0} = \{\text{id}\}.$$

Case 2: $k + l > 0$.

We denote by Γ'_{M_0} the group of all

$$M = (M_1, \ldots, M_n) \in \Gamma$$

$$M_i = \begin{pmatrix} \alpha_i & 0 \\ 0 & \alpha_i^{-1} \end{pmatrix}; \quad \alpha_i > 0, \quad 1 \le i \le k,$$

$$M_i = \begin{pmatrix} 1 & b_i \\ 0 & 1 \end{pmatrix}; \quad k < i \le l,$$

$$M_i \in SO(2,\mathbf{R}); \quad l < i \le n.$$

This is a subgroup of finite index of Γ_{M_0}.

We may identify Γ'_{M_0} with a discrete subgroup of

$$\mathbf{R}^{k+l} \times SO(2,\mathbf{R})^{n-k-l}$$

by means of the imbedding

$$M \mapsto (\log \alpha_1, \ldots, \log \alpha_k, b_{k+1}, \ldots, b_l, M_{l+1}, \ldots, M_n).$$

The projection

$$\mathbf{R}^{k+l} \times SO(2, \mathbf{R})^{n-k-l} \to \mathbf{R}^{k+l}$$

is a proper mapping, because $SO(2, \mathbf{R})$ is compact. Hence the image of Γ'_{M_0} in $\mathbf{R}^{k+l}$ is a discrete subgroup

$$L \subset \mathbf{R}^{k+l}.$$

We denote by

$$P \subset \mathbf{R}^{k+l}$$

a fundamental domain of L. From our irreducibility assumption we obtain that the projection defines a bijection

$$\Gamma'_{M_0} \xrightarrow{\sim} L \quad (k + l > 0).$$

We now obtain that a fundamental domain of Γ'_{M_0} is defined by:

$$\mathcal{F}(M_0) = \{z = x + iy \in \mathbf{H}^n, \quad (\log y_1, \ldots, \log y_k, x_{k+1}, \ldots, x_l) \in P\}$$

We notice that we have no restriction for
a) the coordinates $x_1, \ldots, x_k$ (hyperbolic components)
b) the coordinates $y_{k+1}, \ldots, y_l$ (parabolic components)
c) $z_{l+1}, \ldots, z_n$ (elliptic components).

Hence we are led to the computation of the following integrals in the one-variable case:

2.3 Lemma. *Assume $n = 1$. We have*

1)
$$\int_{-\infty}^{\infty} k(M_0, z) \, dx = 0$$

if

$$M_0 = \begin{pmatrix} \alpha & 0 \\ 0 & \alpha^{-1} \end{pmatrix}, \quad \alpha \neq \pm 1$$

(hyperbolic case).

2)
$$\int_0^{\infty} k(M_0, z) \frac{dy}{y^2} = 2i \frac{1}{2r - 1} \beta^{-1}$$

if

$$M_0 = \begin{pmatrix} 1 & \beta \\ 0 & 1 \end{pmatrix}, \quad \beta \neq 0$$

(parabolic case).

3)
$$\int_H k(M_0, z) \frac{dx\,dy}{y^2} = \frac{4\pi}{2r - 1} \frac{\zeta^r}{1 - \zeta}$$

if M_0 is elliptic with rotation factor ζ (If a is the unique elliptic fixed point of M_0, then M_0 has the form

$$w \to \zeta w$$

in the coordinates $w = (z - a)(z - \bar{a})^{-1} \in E$).

Proof.
1) We have

$$k(M_0, z) = \left(\frac{a^2 z - \bar{z}}{z - \bar{z}}\right)^{-2r} a^{2r}.$$

Therefore the integral is of the type

$$\int_{-\infty}^{\infty} \frac{dx}{(x + a)^{2r}}, \quad a \notin \mathbf{R},$$

which, by the residue theorem, equals 0.

2) The integral

$$\int_0^{\infty} k(M_0, z)\,dy/y^2 = \int_0^{\infty} y^{2r-2}(y + \beta/2i)^{-2r}\,dy$$

can be computed by means of partial integration (differentiation of the first and integration of the second factor of the integrand).

3) We express the integral by means of polar coordinates

$$w = (z - a)(z - \bar{a})^{-1} = \rho e^{i\varphi}$$

of the unit disc. We recall from §1 that the invariant measure transforms like

$$\frac{dx\,dy}{y^2} \mapsto \frac{4\rho\,d\rho\,d\varphi}{(1 - \rho^2)^2}.$$

A simple computation now yields

$$\int_H k(M_0, z)\frac{dx\,dy}{y^2} = 4 \int_0^{2\pi} \int_0^1 \frac{(1 - \rho^2)^{2r-2}}{(1 - \zeta\rho^2)^{2r}} \zeta^r \rho\,d\rho\,d\varphi$$

$$= 4\pi\zeta^r \int_0^1 \frac{(1 - t)^{2r-2}}{(1 - \zeta t)^{2r}}\,dt = \frac{4\pi}{2r - 1}\frac{\zeta^r}{1 - \zeta},$$

because, if

$$G(t) := \frac{1}{(2r-1)(\zeta-1)} \left(\frac{1-t}{1-\zeta t} \right)^{2r-1} ,$$

then

$$G'(t) = \frac{(1-t)^{2r-2}}{(1-\zeta t)^{2r}} .$$ □

2.4 Lemma. *Assume that $M_0 \in \Gamma$ has no fixed point in $\mathbf{H}^n$. Then the contribution of the conjugacy class $[M_0]$ in the trace formula is 0:*

$$\int_{\mathbf{H}^n / \Gamma_{M_0}} f(M_0, z) \, d\omega = 0 .$$

Proof. We may assume that M_0 has the form described above. The assumption of the lemma is $k + l > 0$.

Case 1: $k > 0$ (hyperbolic components do occur: The assertion follows from 2.2,1).

Case 2: $k = 0$ (no hyperbolic, but parabolic as well as elliptic components might occur. We are going to show that this cannot happen).

Case 2a: $l < n$ (elliptic components do exist):

We first notice that the kernel function $k(M_0, z)$ does not depend on x at the parabolic components. We hence obtain that the integral

$$\int_P dx_1 \cdots dx_l$$

converges. This means that the discrete subgroup $L \subset \mathbf{R}^l$ is a lattice. We especially have isomorphisms

$$\Gamma'_{M_0} \xrightarrow{\ \sim\ } L \cong \mathbf{Z}^l .$$

Remark. *Two different elements M, N of Γ'_{M_0} are not conjugate in Γ.*

Proof. The last components of M, N are contained in $SO(2, \mathbf{R})$. But two elements of $SO(2, \mathbf{R})$ are conjugate in $SL(2, \mathbf{R})$ if and only if they are equal (look at the fixed point i). From our irreducibility assumption we obtain

$$M_n = N_n \quad \Rightarrow \quad M = N .$$ □

We also have

$$\Gamma'_M = \Gamma'_{M_0} \text{ for } M \in \Gamma_{M_0}, \ M \neq E .$$

From the computation of the integrals 2.3 we now obtain for $M \in \Gamma'_{M_0}$, $M \neq E$:

$$\left| \int_{\mathcal{F}(M)} k(M, z) \, d\omega \right| \geq \delta \cdot |\beta_1 \cdot \ldots \cdot \beta_l|^{-1} ,$$

where $\delta > 0$ is independent of M. (The contributions of the elliptic components can be packed into the constant δ). We now obtain that the series

$$\sum_{\beta \in L - \{0\}} |\beta_1 \cdots \beta_l|^{-1}$$

converges, which is contradiction, i.e. the case 2a cannot occur.

Case 2b: (all the components of M_0 are parabolic):
As in the case 2a, we see that the discrete subgroup $L \subset \mathbf{R}^n$ is a lattice, now of rank n. The difference to the case 2a now is, that two different elements

$$M = \begin{pmatrix} 1 & \beta \\ 0 & 1 \end{pmatrix} , \quad N = \begin{pmatrix} 1 & \tilde{\beta} \\ 0 & 1 \end{pmatrix}$$

may be conjugate:

$$P M P^{-1} = N .$$

The matrix P is necessarily upper triangular. We obtain

$$\tilde{\beta} = \varepsilon \beta$$

where ε is a multiplier of the translation lattice L,

$$\varepsilon L = L , \quad \varepsilon > 0 ,$$

especially

$$N\varepsilon = 1 .$$

We recall that the group of all such multipliers is a discrete subgroup

$$\Lambda \subset (\mathbf{R}_+)^n$$

of rank $\leq n - 1$. As in the second case, we may conclude that the sum

$$\sideset{}{'}\sum_{a \in L \bmod \Lambda} |Na|^{-1}$$

converges. The summation is now taken over a complete system of elements $a \in L$, $a \neq 0$, which are not associate mod Λ. But it is easy to show by comparison with an integral that this series cannot converge.   □

We have proved that only the elements of finite order in Γ give a non-zero contribution. The identity gives the main term

$$\mathrm{vol}(\mathbf{H}^n/\Gamma)(2r - 1)^n .$$

2.5 Lemma. *Let a run over a system of representatives of Γ-equivalence classes of elliptic fixed points. Denote by $Z \subset \Gamma$ the l elements which define the identity transformation. Then*

$$\bigcup_a (\Gamma_a - Z)$$

is a set of representatives of Γ-conjugacy classes of elements of finite order not contained in Z.

The proof is easy and can be left to the reader.

We finally obtain

2.6 Theorem. *Let $\Gamma \subset SL(2,\mathbf{R})^n$ be a discrete subgroup such that $\mathbf{H}^n/\Gamma$ is compact. Assume that Γ is irreducible (I2.13). Then for $r > 1$ we have*

$$\dim[\Gamma, 2(r,\dots,r)] = \mathrm{vol}(\mathbf{H}^n/\Gamma)(2r-1)^n + \sum_a E_r(\Gamma, a) ,$$

where a runs over a set of representatives of Γ-classes of elliptic fixed points with

$$E_r(\Gamma, a) = \frac{1}{\#\Gamma_a} \sum_{\substack{M \in \Gamma_a \\ M \text{ not the identity}}} N \frac{\zeta^r}{1-\zeta} ,$$

where ζ denotes the rotation factors of M.

§3 The Contribution of the Cusps to the Trace Formula

In the following $\Gamma \subset SL(2,\mathbf{R})^n$ denotes a discrete subgroup such that the extended quotient $(\mathbf{H}^n)^*/\Gamma$ is compact. We assume that Γ has cusps, which means that $\mathbf{H}^n/\Gamma$ is not compact. We also assume the first condition of irreducibility, i.e. the n projections

$$\Gamma \to SL(2,\mathbf{R})$$

are injective.

We choose a precise fundamental domain of the form described in Chap. I, §2: Let

$$\kappa_1, \dots, \kappa_h$$

be a system of representatives of the cusp classes. We may choose cusp sectors

$$V_1, \dots, V_h$$

and a set $K \subset \mathbf{H}^n$ which is contained in a compact subset of $\mathbf{H}^n$ such that the union

$$F = K \cup V_1 \cup \cdots \cup V_h$$

is disjoint and such that F is a (precise) fundamental domain of Γ.

We now assume that ∞ is a cusp of Γ and denote by V_∞ a cusp sector at infinity. We also denote by $\Gamma_\infty^{(1)}$

$$\Gamma_\infty^{(1)} \subset \Gamma_\infty \subset \Gamma$$

the subgroup of all translations, i.e. the set of all elements

$$M = (M_1, \ldots, M_n); \quad M_j = \pm \begin{pmatrix} 1 & \alpha_j \\ 0 & 1 \end{pmatrix}.$$

The vector $\alpha = (\alpha_1, \ldots, \alpha_n)$ is an element of the translation lattice

$$\mathbf{t} = \mathbf{t}(\Gamma).$$

The mapping

$$\Gamma_\infty^{(1)} \to \mathbf{t}, \quad M \mapsto \alpha$$

is a homomorphism with finite kernel. The factor group $\Gamma_\infty / \Gamma_\infty^{(1)}$ is isomorphic to the group Λ of multipliers. The isomorphism is given by

$$M \mapsto \varepsilon, \quad \text{where} \quad Mz = \varepsilon z + \alpha.$$

3.1 Proposition. *The series*

$$\sum_{M \in \Gamma - \Gamma_\infty} |k(M, z)|$$

is bounded on V_∞.

(Recall:

$$k(M, z) = \left[\frac{k(Mz, z)}{k(z, z)} \right]^r j(M, z)^r \quad)$$

3.1₁ Corollary. *The integral*

$$\int_{V_\infty} \sum_{M \in \Gamma - \Gamma_\infty} k(M, z) \, d\omega_z$$

is termwise integrable.

The corollary follows from the Lebesgue limit theorem (AII.5).
For the proof of 3.1 we need the following Lemmata $3.1_2 - 3.1_5$.

3.1₂ Lemma. *The series*

$$\sum_{\alpha \in t} N(1 + (\alpha + x)^2)^{-s}$$

converges uniformly in $x \in \mathbf{R}^n$ *if* $s > 1/2$ *(Here 1 denotes the vector* $(1, \ldots, 1)$*)*.

Proof. The series depends only on $x \mod t$. We may hence assume that the components of x are bounded by a certain constant C. An elementary inequality states: There exists an $\varepsilon = \varepsilon(C) > 0$ such that

$$1 + (\alpha + x)^2 \geq \varepsilon(1 + \alpha^2)$$

for all

$$\alpha \in \mathbf{R} \quad \text{and} \quad x \in \mathbf{R}, \quad |x| \leq C.$$

The inequality shows, that the series in 3.1₂ can be estimated up to a constant factor by the integral

$$\int_{\mathbf{R}^n} N(1 + x^2)^{-s} \, dx_1 \cdots dx_n = \left[\int_{-\infty}^{\infty} (1 + x^2)^{-s} \, dx \right]^n. \qquad \square$$

Before we state the next lemma we recall that the three conditions

$$M = \begin{pmatrix} a & b \\ c & d \end{pmatrix} \in \Gamma_\infty,$$

$$c = 0,$$

$$Nc = 0$$

are equivalent. We also notice that the expression $|Nc|$ does not change if one multiplies M from the left by an element of Γ_∞ or from the right by an element of $\Gamma_\infty^{(1)}$:

$$\pm \begin{pmatrix} a & b \\ c & d \end{pmatrix} \begin{pmatrix} 1 & \alpha \\ 0 & 1 \end{pmatrix} = \pm \begin{pmatrix} a & a\alpha + b \\ c & c\alpha + d \end{pmatrix}.$$

3.1₃ Lemma. *The series*

$$\sum_{M \in \Gamma_\infty \backslash (\Gamma - \Gamma_\infty) / \Gamma_\infty^{(1)}} |Nc|^{-s}$$

converges if $s > 2$.

Proof. We make use of the fact that the series (I.5.7)

$$\sum_{M \in \Gamma_\infty \backslash \Gamma} N(c^2 + d^2)^{-s/2} \qquad (s > 2)$$

is convergent. We now divide Γ into double cosets

$$\Gamma_\infty \backslash \Gamma / \Gamma_\infty^{(1)}$$

and obtain for the sum the expression

$$\sum_{M \in \Gamma_\infty \backslash \Gamma / \Gamma_\infty^{(1)}} \sum_{\Gamma_\infty^{(1)}} N(c^2 + (c\alpha + d)^2)^{-s/2} .$$

Here α denotes the translation vector of the corresponding translation matrix in $\Gamma_\infty^{(1)}$. If $M \notin \Gamma_\infty$ we may write

$$\sum_{\Gamma_\infty^{(1)}} N(c^2 + (c\alpha + d)^2)^{-s/2} = |Nc|^{-s} \sum_{\Gamma_\infty^{(1)}} N(1 + \left(\alpha + \frac{d}{c}\right)^2)^{-s/2} .$$

But the series on the right hand side has a positive lower bound by 3.1_2. $\square$

3.1$_4$ Lemma. *Choose*

$$c \in \mathbf{R}^n , \quad Nc \neq 0 .$$

There exists a constant C depending only on Λ and s such that

$$\sum_{e \in \Lambda} N(\epsilon c^2 + 1)^{-s} \leq C \cdot |Nc|^{-s}$$

if $s > 0$.

Proof. We want to estimate the series by means of an integral. For this purpose we consider the function

$$f(t_1, \ldots, t_{n-1}) = N(c^2 t + 1)^{-s} ,$$

where

$$t = (t_1, \ldots, t_n) \in (\mathbf{R}_+)^n , \quad Nt = 1 .$$

Let $K \subset (\mathbf{R}_+)^{n-1}$ be any compact subset. Then there exists a constant C_1 depending only on s such that

$$f(t_1, \ldots, t_{n-1}) \leq C_1 f(\lambda_1 t_1, \ldots, \lambda_{n-1} t_{n-1})$$

if

$$(\lambda_1, \ldots, \lambda_{n-1}) \in K .$$

This follows from the trivial estimation

$$f(t_1, \ldots, t_{n-1}) \leq f(t_1', \ldots, t_{n-1}') \cdot \left(\max_{1 \leq i \leq n} \left(1, \frac{t_i'}{t_i}\right) \right)^{\frac{n}{s}} .$$

We now make use of the fact that the mapping

$$\Lambda \hookrightarrow \mathbf{R}_+^{n-1}$$
$$\varepsilon \mapsto (\varepsilon_1, \ldots, \varepsilon_{n-1})$$

defines an imbedding of Λ as discrete subgroup with compact quotient. Let K be a compact fundamental domain of Λ. In the usual way one can interpret the sum

$$\sum_{\varepsilon \in \Lambda} f(\varepsilon_1, \ldots, \varepsilon_{n-1})$$

as an integral with respect to the invariant measure $(dt_1/t_1) \cdots (dt_{n-1}/t_{n-1})$ along a function which is constant on the (multiplicative) translates of K. The above consideration gives us the existence of a constant C_2 – independent of c – such that

$$\sum_{\varepsilon \in \Lambda} N(c^2\varepsilon + 1)^{-s} = \sum_{\varepsilon \in \Lambda} f(\varepsilon_1, \ldots, \varepsilon_{n-1})$$

$$\leq C_2 \int_{(\mathbf{R}_+)^{n-1}} f(t_1, \ldots, t_{n-1}) \frac{dt_1}{t_1} \cdots \frac{dt_{n-1}}{t_{n-1}} .$$

From $Nt = 1$ we obtain

$$f(t_1, \ldots, t_{n-1}) = (c_1^2 t_1^{1/2} + t_1^{-1/2})^{-s} \cdots (c_n^2 t_n^{1/2} + t_n^{-1/2})^{-s} .$$

Together with the inequality

$$c_n^2 t_n^{1/2} + t_n^{-1/2} \geq 2|c_n|$$

we obtain $(s > 0)$

$$\sum_{\varepsilon \in \Lambda} N(c^2\varepsilon + 1)^{-s} \leq C_3 |c_n|^{-s} \cdot \prod_{i=1}^{n-1} \int_0^\infty (c_i^2 t_i^{1/2} + t_i^{-1/2})^{-s} \, dt_i/t_i .$$

The transformation $t_i \to |c_i|^{-1} t_1$ shows that the value of the i-th integral is $|c_i|^{-s}$ up to a constant factor. $\square$

For the next lemma we notice that the expression

$$Nc^2 \, N(c^2 + 1)^s$$

does not change if one multiplies the matrix $M = \begin{pmatrix} a & b \\ c & d \end{pmatrix} \in \Gamma$ from the left or from the right by an element of $\Gamma_\infty^{(1)}$.

3.1₅ Lemma. *The series*

$$\sum_{M\in\Gamma_\infty^{(1)}\backslash(\Gamma-\Gamma_\infty)/\Gamma_\infty^{(1)}} (Nc)^{-2}N(c^2+1)^{-s}$$

is convergent if $s > 0$.

Proof. Let N be a matrix in Γ_∞ with multiplier ε, i.e.

$$Nz = \varepsilon z + \alpha.$$

Multiplication of $M = \begin{pmatrix} a & b \\ c & d \end{pmatrix}$ from the right with N has the effect

$$c^2 \mapsto c^2\varepsilon.$$

The series of Lemma 3.1₅ hence equals

$$\sum_{M\in\Gamma_\infty\backslash(\Gamma-\Gamma_\infty)/\Gamma_\infty^{(1)}} (Nc)^{-2}\left(\sum_{\varepsilon\in\Lambda} N(c^2\varepsilon+1)^{-s}\right) \quad (s > 0),$$

where Λ denotes the group of multipliers of Γ. Now the proof of Lemma 3.1₄ follows from the Lemmata 3.1₃ and 3.1₄. $\square$

Proof of Proposition 3.1.
We have

$$S(z) := \sum_{M\in\Gamma-\Gamma_\infty} |k(M,z)| = \sum_{M\in\Gamma_\infty^{(1)}\backslash(\Gamma-\Gamma_\infty)} \left|\frac{j(M,z)^r}{k(z,z)^r}\right| \cdot \sum_{\Gamma_\infty^{(1)}} |k(Mz+\alpha,z)|^r,$$

where α denotes the translation vector of the corresponding matrix in $\Gamma_\infty^{(1)}$. We first estimate the inner sum which equals up to a constant factor

$$\sum_{\alpha\in t} |N(Mz-\bar{z}+\alpha)|^{-2r} = \sum_{\alpha\in t} N(\eta^2+(\alpha+\beta)^2)^{-r},$$

where

$$\eta = \mathrm{Im}(Mz-\bar{z})$$
$$\beta = \mathrm{Re}(Mz-\bar{z}).$$

For the proof of 3.1 we may assume $y_i \geq 1$ ($1 \leq i \leq n$), because the complement in V_∞ is contained in a compact subset of H^n. We then have

$$\eta_i \geq 1 \quad (1 \leq i \leq n).$$

The elementary inequality

$$(t^2+x^2)^{-r} \leq (t^2+x'^2)^{-r}e^{r|x-x'|},$$

where
$$t \geq 1; \quad x, x' \in \mathbf{R}, \quad r > 0,$$

shows that we may estimate the above sum up to a constant factor which is independent of β and η by the integral

$$\int_{\mathbf{R}^n} |N(\eta^2 + x^2)|^{-r} dx_1 \cdots dx_n .$$

This integral can be computed by means of the transformation $x \to \eta x$. The result is
$$N\eta^{-2r+1}$$

times a constant factor. Thus we have proved

$$S(z) \leq C_1 \sum_{M \in \Gamma_\infty^{(1)} \backslash (\Gamma - \Gamma_\infty)} N\left[(\text{Im}(Mz - \bar{z}))^{-2r+1} \frac{|cz + d|^{-2r}}{y^{-2r}} \right]$$

$$= C_1 \sum_{M \in \Gamma_\infty^{(1)} \backslash (\Gamma - \Gamma_\infty)} N\left[y(|cz + d|^{-2} + 1)^{-2r+1} |cz + d|^{-2r} \right]$$

$$\leq C_1 \sum_{M \in \Gamma_\infty^{(1)} \backslash (\Gamma - \Gamma_\infty)} \frac{Ny}{N(|cz + d|^2 + 1)^r}$$

with a certain constant C_1. We now divide into cosets of $\Gamma_\infty^{(1)}$ from the right and obtain

$$S(z) \leq C_1 \cdot Ny \sum_{M \in \Gamma_\infty^{(1)} \backslash (\Gamma - \Gamma_\infty) / \Gamma_\infty^{(1)}} Nc^{-2r}$$
$$\sum_{\Gamma_\infty^{(1)}} N((x + c^{-1}d + \alpha)^2 + c^{-2}(1 + c^2 y^2))^{-r} ,$$

where α denotes the translation vector of the corresponding matrix in $\Gamma_\infty^{(1)}$. Repeating the same argument as above we obtain

$$S(z) \leq C_2 \sum_{M \in \Gamma_\infty^{(1)} \backslash (\Gamma - \Gamma_\infty) / \Gamma_\infty^{(1)}} \frac{Ny}{|Nc| N(c^2 y^2 + 1)^{r-1/2}}$$

with a constant C_2. An elementary inequality finally gives us

$$S(z) \leq C_2 \sum_{M \in \Gamma_\infty^{(1)} \backslash (\Gamma - \Gamma_\infty) / \Gamma_\infty^{(1)}} Nc^{-2} N(c^2 + 1)^{1-r} .$$

This series is actually convergent for $r > 1$ by Lemma 3.1$_5$. We have thus proved Proposition 3.1. $\qquad \square$

3.2 Proposition. *On V_∞ there exists an estimation*

$$\sum_{M \in \Gamma_\infty} |k(M, z)| < CNy ,$$

where the constant C only depends on Γ_∞ and V_∞ (but not on z).

3.2₁ Corollary. *The function*

$$\sum_{M \in \Gamma_\infty} |k(M, z)|(Ny)^{-s} , \quad s > 0 ,$$

is integrable along V_∞ (with respect to the invariant volume element $d\omega_z$).

3.2₂ Corollary. *The function*

$$\sum_{M \in \Gamma_\infty} k(M, z)$$

is bounded on V_∞ (which follows from 3.1 and 1.3), and we have

$$\int_{V_\infty} \sum_{M \in \Gamma_\infty} k(M, z) \, d\omega = \lim_{s \to 0+} \int_{V_\infty} (Ny)^{-s} \sum_{M \in \Gamma_\infty} k(M, z) \, d\omega .$$

By the first corollary this equals

$$\lim_{s \to 0+} \sum_{M \in \Gamma_\infty} \int_{V_\infty} k(M, z)(Ny)^{-s} \, d\omega .$$

Again the corollary follows from the Lebesgue limit theorem (AII.5).

Proof. We have

$$\sum_{M \in \Gamma_\infty} |k(M, z)| = \sum_{M = \left(\begin{smallmatrix} a & b \\ 0 & d \end{smallmatrix}\right) \in \Gamma_\infty} Ny^{2r} |N(d^{-1}(az + b) - \bar{z})|^{-2r} .$$

(The quotient $\varepsilon = ad^{-1}$ is a multiplier, hence $N\varepsilon = 1$. From $ad = 1$ we obtain $Na = Nd = \pm 1$.) We now divide Γ_∞ into cosets of $\Gamma_\infty^{(1)}$ from the right and obtain

$$\sum_{M \in \Gamma_\infty / \Gamma_\infty^{(1)}} (Ny)^{2r} \sum_{\Gamma_\infty^{(1)}} |N(d^{-1}(az + a\alpha + b) - \bar{z})|^{-2r} =$$

$$\sum_{M \in \Gamma_\infty / \Gamma_\infty^{(1)}} (Ny)^{2r} \sum_{\Gamma_\infty^{(1)}} |N[(1 + \varepsilon^{-1})^2 y^2 + (\alpha + x - \varepsilon^{-1}x + a^{-1}b)^2]|^{-r} ,$$

where α denotes the translation vector of the corresponding matrix in $\Gamma_\infty^{(1)}$. The sum over α can be estimated by an integral like that described in the proof of 3.1. The result is

$$\sum_{M\in\Gamma_\infty} |k(M,z)| \leq CNy \sum_{\Gamma_\infty/\Gamma_\infty^{(1)}} N(1+\varepsilon)^{1-2r}.$$

The sum on the right hand side is convergent by Lemma 3.1₄. □

By means of Propositions 3.1 and 3.2 we are now going to express the rank formula as a sum of contributions of the conjugacy classes of Γ. First we choose transformations which transform the given set of representatives of the cusp classes to infinity

$$N_j\kappa_j = \infty \qquad (1 \leq j \leq h).$$

We may assume that there is a (large) constant C such that V_j is a fundamental domain of Γ_{κ_j} in

$$U_j = N_j^{-1}U_C$$
$$U_C = \{z \in H^n; Ny > C\}$$

(Recall:

$$F = K \cup V_1 \cup \cdots \cup V_h$$

is a precise fundamental domain of Γ.)

We denote by S_j $(1 \leq j \leq h)$ the set of all elements of the stabilizer Γ_{κ_j} which are different from the identity (as transformations) and by

$$S = S_1 \cup \cdots \cup S_h.$$

We now split the trace formula

$$l \cdot \left(\frac{2r-1}{4\pi}\right)^{-n} \dim[\Gamma, 2r]_0 = \int_F \sum_{M\in\Gamma} k(M,z)\,d\omega = A + B,$$

where

$$A = \int_F \sum_{M\in\Gamma-S} k(M,z)\,d\omega$$
$$B = \int_F \sum_{M\in S} k(M,z)\,d\omega.$$

In the first integral we may interchange summation and integration. In the following M_0 runs over a complete set of representatives of the conjugacy classes of Γ. We have

$$\Gamma - S = \bigcup_{M_0}[M_0]',$$

where

$$[M_0]' = [M_0] \cap (\Gamma - S)$$
$$([M_0] = \{N M_0 N^{-1}, \ N \in \Gamma\}).$$

For the first integral we obtain

$$A = \sum_{M_0} \sum_{M \in [M_0]'} \int_F k(M, z)\, d\omega = \sum_{M_0} \int_{F(M_0)'} k(M_0, z)\, d\omega,$$

where

$$F(M_0)' = \bigcup_{\substack{M M_0 M^{-1} \notin S \\ M \ \mathrm{mod} \ \Gamma_{M_0}}} M^{-1}(F).$$

Here " mod Γ_{M_0} " means that M runs over a fixed system of representatives with respect to the equivalence relation

$$M \sim N \quad \Longleftrightarrow \quad M M_0 M^{-1} = N M_0 N^{-1}.$$

The domain $F(M_0)'$ is part of the fundamental domain

$$F(M_0) = \bigcup_{M \in [M_0], \ M \ \mathrm{mod} \ \Gamma_{M_0}} M^{-1}(F)$$

of the centralizer Γ_{M_0} of M_0 in Γ. In the special case

$$[M_0] \subset \Gamma - S$$

we have of course

$$F(M_0) = F(M_0)'.$$

We now treat the second integral

$$B = \sum_{j=1}^{h} B_j, \quad B_j = B_j' + B_j'',$$

where

$$B_j' = \int_{F - V_j} \sum_{M \in S_j} k(M, z)\, d\omega$$

$$B_j'' = \int_{V_j} \sum_{M \in S_j} k(M, z)\, d\omega.$$

In the integral B_j' summation and integration can again be interchanged

$$B_j' = \sum_{M_0} \sum_{M \in [M_0] \cap S_j} \int_{F - V_j} k(M, z)\, d\omega = \sum_{M_0} \int_{F_j(M_0)} k(M_0, z)\, d\omega,$$

where

$$F_j(M_0) := \bigcup_{\substack{M M_0 M^{-1} \in S_j, \\ M \bmod \Gamma_{M_0}}} M^{-1}(F - V_j).$$

We obtain

$$A + B_1' + \cdots + B_h' = \sum_{M_0} \int_{F(M_0)^*} k(M_0, z) \, d\omega,$$

where

$$F(M_0)^* = F(M_0)' \cup \bigcup_{j=1}^{h} F_j(M_0) = F(M_0) - \bigcup_{j=1}^{h} \bigcup_{\substack{M M_0 M^{-1} \in S_j, \\ M \bmod \Gamma_{M_0}}} M^{-1}(V_j).$$

In the remaining integrals B_j'' we have to introduce convergence generating factors. If we apply Proposition 3.2 to the conjugate group $N_j \Gamma N_j^{-1}$ instead of Γ, then after a simple transformation we obtain

$$B_j'' = \lim_{s \to 0+} \sum_{M \in S_j} \int_{V_j} \frac{k(M, z) \, d\omega}{(Ny)^s |j(N_j, z)|^s}$$

$$= \lim_{s \to 0+} \sum_{M_0} \sum_{\substack{M M_0 M^{-1} \in S_j, \\ M \bmod \Gamma_{M_0}}} \int_{M^{-1}(V_j)} \frac{k(M_0, z) \, d\omega}{(Ny)^s |j(N_j M, z)|^s}.$$

3.3 Proposition. *Let M_0 run over a complete system of representatives of the conjugacy classes of Γ. For each M_0 we select a fixed set of representatives of the cosets $\Gamma_{M_0} \backslash \Gamma$, where Γ_{M_0} denotes the centralizer of M_0 in Γ. Then the trace formula may be written as follows*

$$l \cdot \left(\frac{2r-1}{4\pi}\right)^{-n} \dim[\Gamma, 2(r, \ldots, r)]_0 = \lim_{s \to 0+} \sum_{M_0} \left\{ \int_{F(M_0)^*} k(M_0, z) \, d\omega \right.$$

$$\left. + \sum_{j=1}^{h} \sum_{\substack{M M_0 M^{-1} \in S_j, \\ M \bmod \Gamma_{M_0}}} \int_{M^{-1}(V_j)} \frac{k(M_0, z) \, d\omega}{(Ny)^s |j(N_j M, z)|^s} \right\},$$

where

$$F(M_0)^* = F(M_0) - \bigcup_{j=1}^{h} \bigcup_{\substack{M M_0 M^{-1} \in S_j, \\ M \bmod \Gamma_{M_0}}} M^{-1}(V_j)$$

and $F(M_0)$ is a certain fundamental domain of Γ_{M_0}.

We call the expression between the big brackets the **contribution** of M_0. We are now going to compute this contribution in several cases.

Case 1. M_0 *is either the identity (as a transformation) or no cusp of Γ is a fixed point of M_0.*

This is the case if, for example, M_0 is of finite order!
In this case we have $F(M_0)^* = F(M_0)$ and the contribution of M_0 reduces to

$$\int_{F(M_0)} k(M_0, z)\, d\omega \ .$$

We may apply the method of the cocompact case (§2) to compute this integral and we will obtain the same formula for this integral as in the cocompact case. The integral especially vanishes if M_0 is not of finite order. If M_0 is of finite order one obtains the same contributions as in the cocompact case.

Case 2. M_0 *is not the identity and fixes two different cusps (especially $n > 1$).*

We will show that the contribution of M_0 vanishes. For this purpose we may conjugate the group Γ and hence assume that ∞ is a cusp of Γ, i.e.

$$M_0 z = \varepsilon_0 z + b \,, \quad \varepsilon_0 \neq (1, \dots, 1) \,.$$

It is easy to see that each component of ε_0 is different from 1. Hence after a further conjugation we may assume that the second fixed point of M_0 is 0, i.e. $b = 0$. We have to determine the centralizer of M_0. Denote by Λ_0 the group of all multipliers ε such that

$$z \mapsto \varepsilon z \quad (\text{not only } z \mapsto \varepsilon z + b_\varepsilon)$$

is contained in Γ. The centralizer of M_0, more precisely the corresponding group of transformations, consists of all

$$z \mapsto \varepsilon z \,, \quad \varepsilon \in \Lambda_0 \,.$$

We now have to investigate the domain $F(M_0)^*$. Let $\kappa_\alpha, \kappa_\beta$ be the two cusps in our (fixed) set of representatives which are equivalent to $\infty, 0$. (Of course $\alpha = \beta$ is possible.)

$$\kappa_\alpha = A(\infty) \,, \quad \kappa_\beta = B(0) \,; \quad A, B \in \Gamma \,.$$

We must determine all $M \in \Gamma$ such that

$$M M_0 M^{-1} \in S,$$

equivalently

$$MM_0M^{-1}\kappa_j = \kappa_j$$

or

$$M_0M^{-1}\kappa_j = M^{-1}\kappa_j$$

for some j.

This is the case if

$$M^{-1}\kappa_j \in \{\infty, 0\}.$$

Our result is

$$MM_0M^{-1} \in S \quad \Longleftrightarrow \quad M \in \Gamma_{\kappa_\alpha}A \text{ or } M \in \Gamma_{\kappa_\beta}B$$

and therefore

$$F(M_0)^* = F(M_0) - (F_\alpha \cup F_\beta),$$

where

$$F_\alpha = \bigcup_{\substack{M \in \Gamma_{\kappa_\alpha}A \\ M \bmod \Gamma_{M_0}}} M^{-1}(V_\alpha)$$

(analogous for F_β).

Notation:

$$E_\alpha = \bigcup_{M \in \Gamma_{\kappa_\alpha}A} M^{-1}(V_\alpha)$$

$$\widetilde{E}_\beta = \bigcup_{M \in \Gamma_{\kappa_\beta}B} M^{-1}(V_\beta).$$

From the definition of the cusp sectors it follows

$$E_\alpha = \{z \in \mathsf{H}^n \mid Ny > C_1\}$$

$$\widetilde{E}_\beta = \{z \in \mathsf{H}^n \mid N(\mathrm{Im}\left(\frac{-1}{z}\right)) > C_2\}$$

with certain constants C_1, C_2. It is no loss of generality to assume $C_1 > 1, C_2 > 1$. We have especially $E_\alpha \cap \widetilde{E}_\beta = \emptyset$ (which is automatically true if $\alpha \neq \beta$). From the description of Γ_{M_0} we see that Γ_{M_0} acts on E_α and $\widetilde{E}_\beta$. This implies

1) F_α (resp. F_β) is a fundamental domain of Γ_{M_0} in E_α (resp. $\widetilde{E}_\beta$)

and as a consequence

2) $F(M_0)^*$ is a fundamental domain of Γ_{M_0} in $\mathsf{H}^n - (E_\alpha \cup \widetilde{E}_\beta)$.

After these preparations we are able to compute the contribution of M_0. In 3.3 we find the expressions

$$j(N_\alpha M, z), \text{ where } M \in \Gamma_{\kappa_\alpha}A$$

and

$$j(N_\beta M, z), \text{ where } M \in \Gamma_{\kappa_\beta} B \quad .$$

In both cases these expressions do not depend on M. For example

$$j(N_\alpha M, z) = j(N_\alpha A A^{-1} M, z) = j(N_\alpha A, A^{-1} M z) \cdot j(A^{-1} M, z) .$$

But $j(A^{-1}M, z) = 1$ since $A^{-1}M$ fixes ∞ and ∞ is a cusp of the conjugate group $A^{-1}\Gamma A$. The contribution of M_0 in the right hand side of the formula in 3.3 is the sum of the following three integrals

1) $\displaystyle\int_{F(M_0)^*} k(M_0, z) \, d\omega$

2) $\displaystyle\int_{F_\alpha} \frac{k(M_0, z) \, d\omega}{(Ny)^s |j(N_\alpha A, A^{-1} M z)|^s}$

3) see 2) but replace α, A with β, B.

In all three cases the integrand is invariant under a transformation

$$z \mapsto Mz \quad (= \varepsilon z \text{ with } N\varepsilon = 1) \text{ for } M \in \Gamma_{M_0} .$$

This is obvious for $k(M_0, z)$ and follows for the occuring j-factors from

$$N_\alpha A(\infty) = \infty \qquad (\Rightarrow j(N_\alpha A, z) = \text{const.}),$$
$$N_\beta B(0) = \infty \qquad (\Rightarrow j(N_\beta B, z) = \text{const.} \cdot Nz^{-2}) .$$

We may now replace $F_\alpha, F_\beta, F(M_0)^*$ by any fundamental domain of Γ_{M_0} in $E_\alpha, E_\beta, H^n - (E_\alpha \cup E_\beta)$.

We are now able to prove that each of the three integrals vanishes. The easiest is the second one. We notice that a fundamental domain of Γ_{M_0} in $E_\alpha = \{z \in H^n, Ny > C_1\}$ can be described by certain conditions on the imaginary part y and no restriction for the real part x of z. So it is sufficient to prove

$$\int_{\mathbf{R}^n} (\varepsilon_0 z - \bar{z})^{-2r} \, dx = 0 .$$

This follows immediately from the residue theorem (compare 2.3.1). The same method can be applied to the third integral after the transformation $z \mapsto -1/z$. The most involved integral is the first one. Here one has to determine a suitable fundamental domain of Γ_{M_0} in the domain

$$\{z \in H^n \mid \quad Ny < C_1, \quad Ny < C_2 |Nz|^2\} .$$

Recall that Γ_{M_0} consists of transvections

$$z \mapsto \varepsilon z, \quad \varepsilon \in \Lambda_0 ,$$

where Λ_0 is a certain discrete subgroup of the norm-one space in $\mathbf{R}_+^n$.

We introduce new coordinates

$$z = \rho e^{i\varphi},$$

where

$$\rho_j > 0, \quad 0 < \varphi_j < \pi \quad (1 \le j \le n).$$

The exponential function has to be taken component-wise. The functional determinant is $N\rho \, d\rho \, d\varphi$, hence

$$d\omega = \frac{dx \, dy}{Ny^2} = N\rho^{-1} \frac{d\rho \, d\varphi}{N(\sin\varphi)^2}.$$

The new inequalities for our domain are

$$C_2^{-1} N \sin\varphi < N\rho < C_1 N (\sin\varphi)^{-1}.$$

The group of multipliers Λ_0 has the effect

$$\rho \mapsto \varepsilon\rho, \quad \varepsilon \in \Lambda_0,$$

and no effect on the variable φ. Hence we may determine a domain $B \subset \mathbf{R}_+^{n-1}$ such that a fundamental domain of Λ_0 is described by

$$(\rho_1, \ldots, \rho_{n-1}) \in B$$

and no condition for ρ_n and φ. Our integrand $k(M_0, z) \, d\omega$ equals

$$(2i)^{2rn} \cdot \frac{N\rho^{-1} \cdot N(\sin\varphi)^{2r-2}}{N(\varepsilon_0 e^{i\varphi} - e^{-i\varphi})^{2r}} \, d\rho \, d\varphi.$$

We first integrate along the ρ-variable. Up to factors independent of ρ we have

$$\int\limits_{\substack{a < N\rho < b \\ (\rho_1, \ldots, \rho_{n-1}) \in B}} (N\rho)^{-1} \, d\rho,$$

where

$$a = C_2^{-1} N \sin\varphi, \quad b = C_1 N (\sin\varphi)^{-1}.$$

To compute the integral we introduce the variables

$$u_1 = \rho_1, \quad \ldots, \quad u_{n-1} = \rho_{n-1}, \quad u_n = N\rho.$$

The integral equals

$$\int_a^b u_n^{-1} \, du_n \int_B \frac{du_1 \cdots du_{n-1}}{u_1 \cdots u_{n-1}}.$$

Notice. From the convergence of the second integral one can conclude that Λ_0 is actually a group of maximal possible rank $n - 1$, hence a subgroup of finite index of the group Λ of all multipliers.

The value of the integral is up to a constant factor

$$\log b - \log a = \log(C_1 C_2 N(\sin\varphi)^{-2}) .$$

To finish the second case it is obviously enough (use $n > 1$) to prove

$$\int_0^\pi \frac{(\sin\varphi)^{2r-2}}{(e^{i\varphi} - \lambda e^{-i\varphi})^{2r}}\, d\varphi = 0 ,$$

where $|\lambda| \neq 1$.

The transformation $\varphi \mapsto \pi - \varphi$ shows that it is sufficient to treat the case $|\lambda| > 1$. The transformation $\varphi \mapsto \varphi + \pi$ shows that we may integrate from 0 to 2π (instead of π). Hence we have to consider the curve integral

$$\oint \frac{(z - z^{-1})^{2r-2} z^{-1}\, dz}{(z - \lambda z^{-1})^{2r}} ,$$

where the path of integration is the border of the unit disc. We may rewrite this integral as

$$\oint \frac{(z^2 - 1)^{2r-2} z\, dz}{(z^2 - \lambda)^{2r}} .$$

The integrand is an analytic function in a domain which contains the closed unit disc (because of $|\lambda| > 1$). The integral vanishes by Cauchy's theorem.

Case 3. *M_0 is not the identity transformation. It fixes a cusp and it has a further fixed point in $\overline{\mathbf{R}}^n$ which is not a cusp.*

We shall see that this case cannot occur. We proceed as in the second case. We can assume that

$$M_0 z = \varepsilon_0 z , \quad \varepsilon_0 \neq (1,\ldots,1)$$

and that ∞ (but not 0) is a cusp of Γ. Again the centralizer of M_0 is given by a discrete subgroup Λ_0 of the norm-one subspace of $\mathbf{R}_+^n$.

The domain $F(M_0)^*$ now is a fundamental domain of Γ_{M_0} in

$$H^n - E_\alpha ; \quad E_\alpha = \{z \in H^n,\ Ny > C_1\} .$$

But in this case the integral

$$\int_{F(M_0)^*} k(M_0, z)\, d\omega$$

is not absolutely convergent, because the integral

$$\int_0^b u_n^{-1} \, du_n \qquad (b > 0)$$

does not exist (compare the computations in the preceding case).

The Contribution of the Parabolic Transformations. A transformation $M_0 \in \Gamma$ is called **parabolic** if its fixed point set in $H^n \cup \overline{R}^n$ consists of exactly one cusp. If this cusp is ∞, M_0 is a translation. Of course, the conjugates of a parabolic element are parabolic again. Before we determine a suitable system of representatives of the parabolic conjugacy classes, we notice some simple facts.

  1) *The fixed points of two conjugate parabolic transformations are equivalent (mod Γ).*

  2) *Two parabolic transformations with the same fixed point κ are conjugate in the group Γ if and only if they are conjugate in the group Γ_κ.*

Recall that we distinguished a system $\{\kappa_1, \ldots, \kappa_h\}$ of representatives of the cusp classes and also transformations which transform them to infinity

$$N_j \kappa_j = \infty \qquad (1 \leq j \leq h).$$

We denote by t_j the translation lattice of $N_j \Gamma N_j^{-1}$ and by Λ_j its group of multipliers. Of course each Γ-conjugacy class of parabolic elements contains a representative which fixes one of the κ_j.

A simple calculation in Γ_∞ (if ∞ is a cusp) now yields

3.4 Lemma. *Let $\mathcal{H}_j$ be a set of representatives of $t_j - \{0\}$ mod Λ_j. Then the set of all*

$$M \in \Gamma_{\kappa_j}, \quad N_j M N_j^{-1}(z) = z + a, \quad a \in \mathcal{H}_j \quad (1 \leq j \leq h),$$

forms a complete system of non-conjugate parabolic elements of Γ.

We now fix one κ_j and investigate the contribution of all $M_0 \in \Gamma_{\kappa_j}$ in our set of representatives. For sake of simplicity we assume $\kappa = \infty$, hence

$$M_0 z = z + a.$$

The centralizer of M_0 is the group of all translations

$$\Gamma_{M_0} = \Gamma_\infty^{(1)}.$$

The same considerations as in the second case show that the contribution of M_0 is

$$\int_{(H^n - E)/\Gamma_\infty^{(1)}} k(M_0, z) \, d\omega + \int_{E/\Gamma_\infty^{(1)}} \frac{k(M_0, z)}{(Ny)^s} \, d\omega,$$

where

$$E = \{z \in H^n \mid Ny > C\}$$

with a certain constant C. This constant does not depend on a. We have

$$k(M_0, z) = \frac{(2i)^{2nr}(Ny)^{2r}}{N(a + 2iy)^{2r}} .$$

Hence the integrands do not depend on x. Therefore the contribution is

$$(2i)^{2nr}\text{vol}(\mathbf{t}) \left[\int_{Ny<C} \frac{Ny^{2r-2}\,dy}{N(a + 2iy)^{2r}} - \int_{Ny>C} \frac{Ny^{2r-2-s}\,dy}{N(a + 2iy)^{2r}} \right] ,$$

where $\text{vol}(\mathbf{t})$ denotes the volume of a fundamental parallelotope of the translation lattice $\mathbf{t}$. We want to compare the expression in the big bracket with

$$\int_{\mathbf{R}^n_+} \frac{Ny^{2r-2-s}\,dy}{N(a + 2iy)^{2r}}$$

and have to estimate the difference. For every $\varepsilon > 0$ there exists an $s_0 = s_0(C) > 0$ such that

$$|Ny^{2r-2} - Ny^{2r-2-s}| \le \varepsilon \quad \text{for} \quad 0 < s \le s_0 .$$

An upper bound for the difference therefore is

$$\varepsilon \int_{\mathbf{R}^n_+} \frac{dy}{N(a^2 + 4y^2)^r} .$$

The transformation $y_i \mapsto a_i y_i \quad (1 \le i \le n)$ shows that this is

$$\varepsilon |Na|^{-2r+1}$$

times a constant. We therefore have proved that the contribution of M_0 may be written as

$$(2i)^{2nr}\text{vol}(\mathbf{t}) \int_{\mathbf{R}^n_+} \frac{Ny^{2r-2-s}\,dy}{N(a + 2iy)^{2r}} + |Na|^{-2r+1}\varphi_a(s)$$

where the error term $\varphi_a(s)$ tends to 0 uniformly in a if $s \to 0^+$. It is well known that the series

$$\sum_{a:(\mathbf{t}-\{0\})/\Lambda} |Na|^{-2r+1}$$

converges (see AI.20). This implies

$$\lim_{s \to 0^+} \sum |Na|^{-2r+1}\varphi_a(s) = 0 .$$

Hence in the final dimension formula (3.3) the term φ_a is neglectible and we may replace the contribution of M_0 by the modified contribution

$$(2i)^{2nr}\mathrm{vol}(t)\prod_{j=1}^{n}\int_0^\infty \frac{y_j^{2r-2-s}dy_j}{(a_j+2iy_j)^{2r}}\,.$$

We investigate the integral

$$\int_0^\infty \frac{y^{2r-2-s}dy}{(a+2iy)^{2r}}$$

where a is a real number different from 0:

1) $a > 0$: By Cauchy's theorem we may deform the path of integration from the right real axis to the lower imaginary axis, where the complex power

$$z^s = e^{s\log z}$$

has to be defined by the main branch of the logarithm which is holomorphic outside the negative real axis including 0 and real for positive real z. The result is that the integral equals

$$(-i)^{2r-1-s}\int_0^\infty \frac{y^{2r-2-s}dy}{(a+2y)^{2r}} = \frac{2^s}{(2i)^{2r-1}}(-i)^{-s}a^{-(1+s)}\int_0^\infty \frac{y^{2r-2-s}dy}{(1+y)^{2r}}\,.$$

2) In the case $a < 0$ one has to deform the path of integration to the upper imaginary axis. An analogous computation shows that the above formula remains valid in the case $a < 0$ if one replaces $(-i)^{-s}a^{-(1+s)}$ by $-i^{-s}|a|^{-(1+s)}$. Uniting both expressions, this may be written as

$$(\mathrm{sgn}\,a)|a|^{-(1+s)}e^{\frac{\pi}{2}is\,\mathrm{sgn}\,a}\,.$$

The transformation $t = (y+1)^{-1}$ finally gives the result

$$\int_0^\infty \frac{y^{2r-2-s}dy}{(a+2iy)^{2r}} = \frac{2^s}{(2i)^{2r-1}}\cdot\mathrm{sgn}\,a\cdot e^{\frac{\pi}{2}is\,\mathrm{sgn}a}|a|^{-(1+s)}\int_0^1 t^s(1-t)^{2r-2-s}\,dt\,.$$

The integral on the right hand side is a usual Beta integral. We only need its limit for $s \to 0^+$, which is $(2r-1)^{-1}$. The modified contribution of M_0 may now be written as

$$\frac{g(s)(2i)^n\mathrm{vol}(t)}{(2r-1)^n}\,\mathrm{sgn}(Na)\,|Na|^{-(s+1)}e^{\frac{\pi}{2}is\,S(\mathrm{sgn}a)}$$

where $g(s)$ is a function independent of a and

$$\lim_{s\to0^+}g(s)=1\,.$$

Of course
$$S(\operatorname{sgn} a) := \operatorname{sgn} a_1 + \ldots + \operatorname{sgn} a_n .$$

We now sum up over a system M_0 of representatives of conjugacy classes of parabolic elements fixing ∞. We will see in the following that the limit

$$\lim_{s \to 0+} \frac{g(s)(2i)^n \operatorname{vol}(t)}{(2r-1)^n} \sum_{M_0} \operatorname{sgn}(Na)|Na|^{-(1+s)} e^{\frac{\pi}{2} i s \, S(\operatorname{sgn} a)}$$

exists. Assuming this for a moment we may call this expression **the contri-bution of our cusp** to the right-hand side in 3.3. Recall that the system of representatives is given by

$$M_0 z = z + a ,$$

where a runs over a system of representatives of $t - \{0\}$ modulo Λ. But each M_0 occurs l times ($l = $ order of the kernel of the natural projection $\Gamma \to (SL(2,\mathbf{R})/\{\pm E\})^n$). Hence the limit will be equal to

$$\frac{l(2i)^n \operatorname{vol}(t)}{(2r-1)^n} \lim_{s \to 0+} \sum_{a:(t-\{0\})/\Lambda} \operatorname{sgn}(Na)|Na|^{-(1+s)} e^{\frac{\pi}{2} i s \, S(\operatorname{sgn} a)} .$$

We treat the cases $n = 1$ and $n > 1$ separately.

1) $n = 1$:   We have
$$t = d\mathbf{Z} , \quad d > 0 .$$

The sum equals

$$d^{-(1+s)} \sum_{n=1}^{\infty} n^{-(1+s)} \left[e^{\frac{\pi}{2} i s} - e^{-\frac{\pi}{2} i s} \right] .$$

Making use of the well-known fact that

$$\lim_{s \to 0+} s \sum_{n=1}^{\infty} n^{-(1+s)} = 1$$

one obtains
$$d^{-1} \cdot \pi i$$

for the limit. Hence the contribution of our cusp is

$$-\frac{2l\pi}{(2r-1)} .$$

2) $n > 1$:   We claim

$$\lim_{s \to 0+} \sum_{a} \operatorname{sgn}(Na)|Na|^{-(1+s)} e^{\frac{\pi}{2} i s \, S(\operatorname{sgn} a)} = \lim_{s \to 0+} \sum_{a} \operatorname{sgn}(Na)|Na|^{-(1+s)} .$$

The existence of the limit on the right hand side is well known (AI.20,21). The limit of the difference (if it exists) is obviously

$$(*) \qquad \frac{\pi}{2}i \lim_{s \to 0+} s \sum_a \operatorname{sgn}(Na)S(\operatorname{sgn} a)|Na|^{-(1+s)} \quad .$$

We now choose a sign vector σ

$$\sigma = (\sigma_1, \ldots, \sigma_n), \quad \sigma_i = \pm 1.$$

From AI.21 we know that the limit

$$A = \lim_{s \to 0+} s \sum_{a\sigma > 0} |Na|^{-(1+s)}$$

exists and is independent of σ. Hence the limit $(*)$ exists and equals

$$\frac{\pi}{2}iA \sum_\sigma \operatorname{sgn}(N\sigma)S(\operatorname{sgn} \sigma),$$

where the sum runs over all 2^n sign vectors. The sum vanishes. We finally obtain

$$\frac{l(2i)^n \operatorname{vol}(t)}{(2r-1)^n} \lim_{s \to 0+} \sum_{a:(t-\{0\})/\Lambda} \operatorname{sgn}(Na)|Na|^{-(1+s)}$$

as contribution of our cusp.

The contribution of the other cusps is obtained by transforming them to ∞ and considering a conjugate group. We are now able to write down the final dimension formula:

Some Notations. Let $t \subset \mathbf{R}^n$ be a lattice and $\Lambda \subset \mathbf{R}^n_+$ a discrete subgroup of multipliers which has maximal rank $n-1$. We define the **Shimizu L-series** by

$$L(t, \Lambda) =$$
$$= \begin{cases} -1/2, & \text{if } n = 1; \\ \dfrac{i^n}{(2\pi)^n}\operatorname{vol}(t) \displaystyle\lim_{s \to 0+} \sum_{a:(t-\{0\})/\Lambda} \operatorname{sgn}(Na)|Na|^{-(1+s)}, & \text{if } n > 1. \end{cases}$$

Here $\operatorname{vol}(t)$ denotes the volume of a fundamental parallelotope of t.

Remark. *Assume $n > 1$ and that there exists a vector*

$$\varepsilon \in \mathbf{R}^n; \quad N\varepsilon = -1, \quad \varepsilon t = t.$$

Then

$$L(t, \Lambda) = 0.$$

This is always the case if n is odd (take $\varepsilon = (-1, \ldots, -1)$).

Remark. *Let $\alpha \in \mathbf{R}_+^n$ be a totally positive vector. Then*

$$L(\mathbf{t}, \Lambda) = L(\alpha \mathbf{t}, \Lambda).$$

Let now $\Gamma \subset SL(2, \mathbf{R})^n$ be a discrete subgroup with cusp κ. We choose a transformation N with $N\kappa = \infty$ and denote by

$$\mathbf{t}_N, \Lambda_N$$

the group of translations, resp. multipliers of the conjugate group $N\Gamma N^{-1}$. By the above remark the value

$$L(\Gamma, \kappa) := L(\mathbf{t}_N, \Lambda_N)$$

does not depend on the choice of N. We can now write down the final formula.

3.5 Theorem. *Assume that $\Gamma \subset SL(2, \mathbf{R})^n$ is a discrete subgroup such that $(\mathbf{H}^n)^*/\Gamma$ is compact. We also assume that the irreducibility condition is satisfied.*

If $r > 1$ we have

$$\dim[\Gamma, 2(r, \ldots, r)]_0 = \operatorname{vol}(\mathbf{H}^n/\Gamma)(2r-1)^n + \sum_a E_r(\Gamma, a) + \sum_\kappa L(\Gamma, \kappa)$$

where a runs over a set of representatives of Γ-classes of elliptic fixed points and κ over a set of representatives of the cusp classes.

The contribution $E(\Gamma, a)$ of an elliptic fixed point is defined as in §2

$$E_r(\Gamma, a) = \frac{1}{\#\Gamma_a} \sum_{\substack{M \in \Gamma_a \\ M \neq \text{identity}}} N \frac{\zeta^r}{1 - \zeta}$$

where ζ is the rotation factor of M, i.e.

$$\frac{Mz - a}{Mz - \overline{a}} = \zeta \frac{z - a}{z - \overline{a}}.$$

The contribution $L(\Gamma, \kappa)$ of a cusp is Shimizu's L-series which we have defined just before.

A Remarkable Symmetry. It follows from Shimizu's formula 3.5 that there exist a natural number r_0 and polynomials in one variable

$$P^{(\nu)}; \quad 0 \le \nu < r_0$$

such that
$$P^{(\nu)}(r) = \dim[\Gamma, 2(\nu + rr_0, \ldots, \nu + rr_0)]_0 \,.$$

The most important is
$$P_\Gamma(r) = P^{(0)}(r) \,,$$

sometimes called Shimizu's polynomial. In the next section there will occur the polynomial
$$Q_\Gamma(r) := P^{(1)}(r) \,,$$

especially its value for $r = 0$.

3.6 Remark. *Assume $n > 1$. Then*

$$P_\Gamma(0) = (-1)^n Q_\Gamma(0) \,.$$

The proof follows from a glance at Shimizu's formula. This symmetry is valid for all three types of contributions. For the main term it is trivial. For the contribution of the elliptic fixed points it follows from

$$\frac{1}{1-\zeta} = -\frac{\zeta^{-1}}{1-\zeta^{-1}}$$

and the fact that with ζ also its inverse ζ^{-1} occurs in the sum defining this contribution.

For the contribution of the cusps it follows from the fact that

$$L(\mathbf{t}, \Lambda) = 0 \qquad \text{if } n \text{ is odd}, \quad n > 1 \,.$$

Final Remark: The values $P_\Gamma(0), Q_\Gamma(0)$ have algebraic geometric meaning. In the next section we will equip $X = (\mathbf{H}^n)^*/\Gamma$ with the structure of a projective algebraic variety. One has

$$\chi(\mathcal{O}_X) = P_\Gamma(0) + h$$
$$\chi(\mathcal{K}_X) = Q_\Gamma(0) + h \,,$$

where $\mathcal{O}_X$ denotes the structure sheaf and $\mathcal{K}_X$ the canonical sheaf in the sense of Serre and χ their Euler-Poincaré characteristic.

The equality
$$\chi(\mathcal{O}_X) - h = (-1)^n (\chi(\mathcal{K}_X) - h) \quad (n > 1)$$
can also be proved by cohomological methods. It follows from the result of the paper [12] and the next section.

§4 An Algebraic Geometric Method

In this section we make use of the fact that the compactification $(\mathbf{H}^n)^*/\Gamma$ (Γ commensurable with a Hilbert modular group) is an algebraic variety. Using the cohomology theory of coherent sheaves, especially the theory of the Hilbert polynomial, we will succeed in expressing $\dim[\Gamma,(2,\ldots,2)]$ by the dimensions of $\dim[\Gamma,(2r,\ldots,2r)]$, $r > 1$, which we computed in the previous sections. Another result will be a formula for the arithmetic genus of a desingularization of $(\mathbf{H}^n)^*/\Gamma$. We do not need the explicit construction of such a desingularization. We have to make use of some results about complex spaces, projective varieties and coherent sheaves, which we cannot develop in this book.

Let $\Gamma \subset SL(2,\mathbf{R})^n$ be a discrete subgroup. We equip the extended quotient

$$X = (\mathbf{H}^n)^*/\Gamma$$

with a certain sheaf of continuous functions: Let

$$V \subset (\mathbf{H}^n)^*/\Gamma$$

be an open subset and

$$U \subset \mathbf{H}^n$$

its inverse image in $\mathbf{H}^n$ with respect to the natural projection. Composing an arbitrary function

$$f : V \to \mathbf{C}$$

with this projection we obtain a function

$$F : U \to \mathbf{C}$$

(which is Γ-invariant). We denote by

$$\mathcal{O}_X(V)$$

the set of all continuous functions f such that F is holomorphic. We obtain a sheaf $\mathcal{O}_X$ of continuous functions and $(X, \mathcal{O}_X)$ is a ringed space.

4.1 Theorem. *Let $\Gamma \subset SL(2,\mathbf{R})^n$ be a discrete subgroup. The extended quotient*

$$X = (\mathbf{H}^n)^*/\Gamma ,$$

equipped with the structure sheaf $\mathcal{O}_X$, is a normal complex space.

Indication of a Proof.

1)   The interior of X, i.e. $\mathbf{H}^n/\Gamma$, is a complex space: What we know is that $\mathbf{H}^n/\Gamma$ locally looks like

$$\mathbf{C}^n/\mathcal{E} ,$$

where $\mathcal{E}$ is a finite group of rotations

$$z \mapsto (\zeta_1 z_1, \ldots, \zeta_n z_n).$$

But it is well known and easily seen that $\mathbf{C}^n/\mathcal{E}$ is a normal algebraic variety (hence a normal complex space).

2) X is a normal complex space at the cusps. Without loss of generality we may assume that the cusp is ∞.

One makes use of the following special case of a

Criterion of Baily and Cartan. Let Y be a locally compact space with a countable basis for its topology. We assume that $a \in Y$ is a given point such that $Y - \{a\}$ is equipped with the structure of a normal complex space. We extend the complex structure to a structure of a ringed space $(Y, \mathcal{O}_Y)$. A function on an open subset $V \subset Y$ is called holomorphic if it is continuous and if its restriction to $V - \{a\}$ is holomorphic with respect to the given complex structure.

Assumptions:

1) There is a fundamental system of open neighbourhoods U of a such that $U - \{a\}$ is connected.
2) The global holomorphic functions $f \in \mathcal{O}_Y(Y)$ separate points (outside a is sufficient).

Then $(Y, \mathcal{O}_Y)$ is a normal complex space.

We apply Cartan's criterion to

$$V = U_C/\Gamma_\infty \cup \{\infty\}$$
$$U_C = \{z \in \mathbf{H}^n \mid Ny > C\}.$$

We consider certain modular forms with respect to the group Γ_∞. They are in fact invariant functions with respect to Γ_∞ and define elements of $\mathcal{O}(V)$! Nevertheless the theory of Poincaré series (Chap. I, §5), especially I.5.3, can be applied to Γ_∞. The separation properties of Poincaré series (which we did not prove completely; see I.5.6 and the concluding remark) show that these functions separate points of U_C/Γ_∞. □

We introduce certain sheaves

$$\mathcal{M}_{2r}, \quad r \in \mathbf{Z}^n,$$

defined on the complex space X. As in the definition of $\mathcal{O}_X$ we denote by $V \subset X$ an open subset and by $U \subset \mathbf{H}^n$ its inverse image. Then $\mathcal{M}_{2r}(V)$ is the set of all holomorphic functions

$$f : U \to \mathbf{C}$$

satisfying the following two conditions:

1) $f(Mz) = \prod_{j=1}^{n}(c_j z_j + d_j)^{2r_j} f(z)$ for all $z \in U$, $M \in \Gamma$.

2) f is regular at the cusps which are contained in U.

What does regularity at a cusp $\kappa \in U$ mean ? If $\kappa = \infty$, then the set U contains

$$U_C = \{z \in H^n \mid Ny > C\}, \quad C \text{ sufficiently large.}$$

The set U_C is invariant with respect to Γ_∞, especially with respect to the translation lattice $\mathbf{t}$. The function f hence admits a Fourier expansion

$$f(z) = \sum_{g \in t^0} a_g e^{2\pi i S(gz)}$$

in U_C. Regularity means

$$a_g \neq 0 \Rightarrow g \geq 0.$$

For an arbitrary cusp $\kappa \in U$ one defines regularity by transforming it to ∞ (compare I4.4). The Götzky-Koecher principle (I.4.9) shows that the regularity condition is automatically satisfied if $n > 1$. The elements of $\mathcal{M}_{2r}(V)$ are so-called **local automorphic forms**. The global sections of the sheaf $\mathcal{M}_{2r}$ are the usual automorphic forms introduced in Chap. I, §4.

$$\mathcal{M}_{2r}(X) = [\Gamma, 2r].$$

The local automorphic form f is called a cusp form, if it vanishes at all the cusps $\kappa \in U$. In case of $\kappa = \infty$ this means

$$a_g = 0 \Rightarrow g > 0,$$

equivalently

$$a_0 = 0.$$

We denote by

$$\mathcal{M}_{2r}^0(V) \subset \mathcal{M}_{2r}(V)$$

the subset of local cusp forms. Obviously $\mathcal{M}_{2r}^0$ is a subsheaf of $\mathcal{M}_{2r}$. Both sheaves are $\mathcal{O}_X$-modules in a natural way.

4.2 Proposition. *The $\mathcal{O}_X$-modules*

$$\mathcal{M}_{2r}, \ \mathcal{M}_{2r}^0$$

are coherent.

Indication of a Proof. We show that $\mathcal{M}_{2r}$ is coherent at the cusps. We restrict to the case of our interest

$$r_1 = \cdots = r_n .$$

If $r \in \mathbf{Z}$ we write

$$\mathcal{M}_{2r} := \mathcal{M}_{2(r,\ldots,r)}$$

The invariance property 1) obviously means for $M \in \Gamma_\infty$ that

$$f(Mz) = f(z)$$

(because the norm of a multiplier is 1). In a neighbourhood

$$U_C/\Gamma_\infty \cup \{\infty\} \hookrightarrow (\mathbf{H}^n)^*/\Gamma \qquad (C \gg 0)$$

this gives us an isomorphism between $\mathcal{M}_{2r}$ and $\mathcal{O}_X$, and $\mathcal{M}_{2r}$ is actually a line bundle close to the cusp.   □

4.3 Lemma. *Let r_0 be a natural number such that the order of each elliptic fixed point* $a \in \mathbf{H}^n$ divides r_0. Then the sheaves*

$$\mathcal{M}_{2r}, \quad r \equiv 0 \bmod r_0 ,$$

are line bundles (i.e. locally isomorphic to $\mathcal{O}_X$).

Notice. If Γ is an irreducible subgroup in sense of I.2.13 and if $(\mathbf{H}^n)^*/\Gamma$ is compact, there exist only finitely many Γ-equivalence classes of points a with non-trivial stabilizer. Hence a number r_0 exists in this case. (This is also true without the irreducibility assumption).

In the following the first assumption of irreducibility I.2.13 and, if $\mathbf{H}^n/\Gamma$ is compact, also the second assumption of irreducibility I.4.12. will, for the sake of simplicity, be considered true.

4.4 Theorem. *Let $\Gamma \subset SL(2,\mathbf{R})^n$ be a discrete subgroup such that the extended quotient $X = (\mathbf{H}^n)^*/\Gamma$ is compact. The line bundles*

$$\mathcal{M}_{2(r,\ldots,r)}, \ r \equiv 0 \bmod r_0 \quad (r_0 \ as \ in \ 4.3) ,$$

are ample for positive r. In particular the complex space X carries a structure as projective algebraic variety.

---

*   Recall that the order of an elliptic fixed point a is the order of the image of Γ_a in $(SL(2,\mathbf{R}/\{\pm E\})^n$.

(Such a structure is unique by the famous comparison theorem of Serre.)

Indication of a Proof: Ampleness means that two different points of X can be separated by a global section of a suitable power of the given line bundle. This can be proved in our case by means of Poincaré series.   □

The Hilbert Polynomial. Let $\mathcal{M}$ be a coherent sheaf on a compact complex space X. The Euler characteristic

$$\chi(\mathcal{M}) := \sum_{j=0}^{\infty} (-1)^j \dim H^j(X, \mathcal{M})$$

is well defined because the cohomology groups are of finite dimension and vanish for sufficiently large j. Let $\mathcal{L}$ be an ample line bundle on X.

1. *There exists a polynomial in one complex variable P such that*

$$\chi(\mathcal{M} \otimes \mathcal{L}^{\otimes r}) = P(r) \,.$$

2. *If r is sufficiently large, $r \gg 0$, we have*

$$H^j(X, \mathcal{M} \otimes \mathcal{L}^{\otimes r}) = 0 \quad \text{if} \quad j > 0 \,,$$

especially

$$\dim(\mathcal{M} \otimes \mathcal{L}^{\otimes r})(X) = P(r) \,, \quad \text{if} \quad r \gg 0 \,.$$

We now apply the theory of the Hilbert polynomial to

$$X = (H^n)^*/\Gamma, \quad \mathcal{M} = \mathcal{M}_2, \quad \mathcal{L} = \mathcal{M}_{2r_0} \,,$$

where r_0 is chosen as in 4.3 and 4.4. We obviously have

$$\mathcal{M} \otimes \mathcal{L}^{\otimes r} = \mathcal{M}_{2+2rr_0} \,.$$

From the theory of the Hilbert polynomial we obtain the existence of a polynomial P with the properties

1) $P(r) = \dim[\Gamma, (2 + 2rr_0, \dots, 2 + 2rr_0)], \ r \gg 0 \,,$

2) $P(0) = \chi(\mathcal{M}_2) \,.$

The polynomial P has actually been computed in the previous sections by means of the trace formula. What we want to compute is the dimension of

$$\mathcal{M}_2(X) \simeq [\Gamma, (2, \dots, 2)] \,.$$

This means that we have to get hold of the cohomology groups of $\mathcal{M}_2$. We want to apply duality theory and for this purpose we consider a desingularization*

$$\pi : \widetilde{X} \to X .$$

Here $\widetilde{X}$ is a nonsingular connected projective variety such that π induces a biholomorphic mapping

$$\pi : \pi^{-1}(X_{\mathrm{reg}}) \to X_{\mathrm{reg}} ,$$

where X_{reg} is the regular locus of X.

We consider on $\widetilde{X}$ the so-called canonical sheaf $\mathcal{K}_{\widetilde{X}}$. The sections of $\mathcal{K}_{\widetilde{X}}$ are holomorphic differential forms of top degree n. In local co-ordinates they have the form

$$f(z)\,dz_1 \wedge \cdots \wedge dz_n \quad (f \text{ holomorphic}) .$$

We now consider the direct images of $\mathcal{K}_{\widetilde{X}}$ on X. An important result of Grauert and Riemenschneider states

4.5 Proposition. *The higher direct images of $\mathcal{K}_{\widetilde{X}}$ on X vanish:*

$$R^p \pi_* \mathcal{K}_{\widetilde{X}} = 0 \quad \text{if} \quad p > 0 .$$

4.5₁ Corollary. $\qquad\qquad \chi(\mathcal{K}_{\widetilde{X}}) = \chi(\pi_* \mathcal{K}_{\widetilde{X}}) .$

We now investigate the direct image $\pi_* \mathcal{K}_{\widetilde{X}}$. Let $V \subset X$ be an open subset and $\widetilde{V} \subset \widetilde{X}$ its inverse image in $\widetilde{X}$. A section $\omega \in (\pi_* \mathcal{K}_{\widetilde{X}})(V)$ is a holomorphic differential form on $\widetilde{V}$. Restricting it to the regular locus of X we obtain a differential form ω_0 on V_{reg}. We denote by U the inverse image of V in $\mathbf{H}^n$ and by U_0 the inverse image of V_{reg} in $\mathbf{H}^n$. The complement $U - U_0$ consists of elliptic fixed points and is hence discrete (by our irreducibility assumption). The pullback of ω_0 to U_0 extends holomorphically to U, because a holomorphic function in more than one variable cannot have isolated singularities (in case $n = 1$ we have $U_0 = U$). The pullback has the form

$$f(z)\,dz_1 \wedge \cdots \wedge dz_n ,$$

---

*   Such a desingularization exists by a general deep result of Hironaka. In the case at hand an explicit desingularization has been constructed by Ehlers [11] generalizing results of Hirzebruch [30] who treated the case $n = 2$. We shall describe this construction in §5.

$$f : U \to \mathbf{C} \quad \text{holomorphic.}$$

The function f is obviously a local automorphic form of weight $(2, \ldots, 2)$. We hence obtain an $\mathcal{O}_X$-linear imbedding

$$\pi_* \mathcal{K}_{\tilde{X}} \subset \mathcal{M}_2$$

and hence may identify $\pi_* \mathcal{K}_{\tilde{X}}$ with a subsheaf of $\mathcal{M}_2$. After this identification we obtain

4.6 Lemma. *We have*

$$\pi_* \mathcal{K}_{\tilde{X}} = \mathcal{M}_2^0 \quad (= \text{sheaf of local cusp forms of weight } (2, \ldots, 2)) \,.$$

Proof. Let f be a local automorphic form of weight $(2, \ldots, 2)$ on some open subset $V \subset X$ and ω_0 the corresponding holomorphic differential form on V_{reg}. We have to show that ω_0 extends holomorphically to the inverse image $\tilde{V}$ of V in $\tilde{X}$ if and only if f is a cusp form.

Remark. *The differential form ω_0 on V_{reg} extends holomorphically to $\tilde{V}$ if and only if for each open subset $W \subset V$ with compact closure in V the integral*

$$\int_{W_{\text{reg}}} \omega_0 \wedge \overline{\omega_0}$$

converges.

Proof of the Remark. The criterion is obviously necessary, because the inverse image $\tilde{W}$ of W in $\tilde{V}$ has compact closure and hence

$$\int_{\tilde{W}} \omega \wedge \overline{\omega}$$

converges.

To prove the sufficiency we first remark that $\tilde{V} - V_{\text{reg}}$ is an analytic subset. As holomorphic functions (hence n-forms) always extend holomorphically over analytic subsets of codimension ≥ 2, we only have to prove the extension into the smooth points of codimension 1. Hence the assertion follows from the following criterion:

Let $E \subset \mathbf{C}$ be the unit disc and f a holomorphic function on $\mathbf{E}^{n-1} \times E - \{0\}$ such that

$$\int_{\mathbf{E}^{n-1} \times \mathbf{E} - \{0\}} |f(z)|^2 \, dv \quad (dv = \text{Euclidean volume element})$$

converges. Then f extends holomorphically to $\mathbf{E}^n$.

This can easily be proved by means of the Laurent expansion in z_n of f. To investigate the convergence of

$$\int_{W_{\text{reg}}} \omega_0 \wedge \overline{\omega_0}$$

we choose a suitable fundamental domain of the inverse image of W in $\mathbf{H}^n$. We know that there is a fundamental domain which consists of the union of a relatively compact subset and a finite number of cusp sectors. It is sufficient to consider the cusp sector at ∞. Now we have the following situation:

On some set $U_C = \{z \in \mathbf{H}^n,\ Ny > 0\}$ the function $f(z)$ is holomorphic and Γ-invariant as well as regular at the cusp ∞. It has to be shown that the integral

$$\int_{U_{C'}/\Gamma_\infty} |f(z)|^2 \, d\omega$$

converges for $C' > 0$ if and only if f vanishes at ∞. This can be proved easily by means of the Fourier expansion of f. This completes the proof of Lemma 4.6. $\square$

We now compare the Euler characteristics of $\mathcal{M}_2$ and $\mathcal{M}_2^0$ by means of the short exact sequence

$$0 \to \mathcal{M}_2^0 \to \mathcal{M}_2 \to \mathcal{M}_2/\mathcal{M}_2^0 \to 0 \,.$$

The sheaf $\mathcal{M}_2/\mathcal{M}_2^0$ vanishes outside the cusps and the stalk at each cusp is isomorphic to $\mathbf{C}$. We obtain

$$\chi(\mathcal{M}_2/\mathcal{M}_2^0) = h = \text{number of cusp classes} \,.$$

The Euler characteristic being additive, we obtain

$$\chi(\mathcal{M}_2) = \chi(\mathcal{M}_2^0) + h \qquad (= \chi(\mathcal{K}_{\tilde{X}}) + h) \,.$$

We now compute $\chi(\mathcal{K}_{\tilde{X}})$. As any nonsingular projective variety carries a Kählerian metric, we may apply Hodge theory (s. App. III). The Hodge numbers of $\tilde{X}$ are

$$h^{p,q} := \dim H^q(\tilde{X}, \Omega^p_{\tilde{X}}) \,,$$

where $\Omega^p_{\tilde{X}}$ denotes the sheaf of holomorphic differential forms of degree p, especially

$$\Omega^n_{\tilde{X}} = \mathcal{K}_{\tilde{X}} \,.$$

We have

$$\chi(\mathcal{K}_{\tilde{X}}) = \sum_{p=0}^{n} (-1)^p h^{n,p} \,.$$

From the duality formulae

$$h^{p,q} = h^{q,p} \; ; \quad h^{p,q} = h^{n-q,n-p}$$

we obtain

$$\chi(\mathcal{K}_{\tilde{X}}) = \sum_{p=0}^{n} (-1)^{n-p} h^{p,0} \,,$$

where

$$h^{p,0} = \dim \Omega^p_{\tilde{X}}(\tilde{X}) \,.$$

4.7 Proposition. *We have*

$$
\begin{aligned}
h^{0,0} &= 1 \,, \\
h^{n,0} &= h^{0,n} = \dim[\Gamma, (2, \ldots, 2)]_0 \,, \\
h^{p,0} &= 0 \quad \text{if} \quad 0 < p < n \,.
\end{aligned}
$$

4.7$_1$ Corollary.

$$\chi(\mathcal{K}_{\tilde{X}}) = (-1)^n + \dim[\Gamma, (2, \ldots, 2)]_0 \,.$$

4.7$_2$ Corollary.

$$\chi(\mathcal{M}_2) = (-1)^n + h + \dim[\Gamma, (2, \ldots, 2)]_0 \,.$$

Proof. We only have to prove the third formula, i.e.: Each holomorphic alternating differential form on $\tilde{X}$ of degree p, $\quad 0 < p < n$, vanishes. Pulling back such a form we obtain a Γ-invariant holomorphic differential form on H^n. But now we may apply the corollary of I.4.11. $\square$

In a later section we shall prove by means of analytic continuation of Eisenstein series

$$\dim[\Gamma, (2, \ldots, 2)]_0 + h = \dim[(2, \ldots, 2)] \quad \text{if} \quad n > 1$$

and

$$\dim[\Gamma, (2, \ldots, 2)]_0 + h - 1 \leq \dim[(2, \ldots, 2)]$$

if Γ is commensurable with a Hilbert modular group. One can show by means of the residue theorem that in the second case we also have an equality ($n = 1$). This gives us

4.7₃ Corollary. *If* Γ *is commensurable with a Hilbert modular group, we have*

$$\chi(\mathcal{M}_2) = (-1)^n + \dim[\Gamma, (2,\ldots,2)] - \varepsilon\,,$$

where

$$\varepsilon = \begin{cases} 1 & \text{if } n = 1 \\ 0 & \text{if } n > 1. \end{cases}$$

The **arithmetic genus** of $\widetilde{X}$ is defined as

$$g := \chi(\mathcal{O}_{\widetilde{X}}) = \text{arithmetic genus}\,.$$

By duality we have

$$\chi(\mathcal{O}_{\widetilde{X}}) = \sum_{p=0}^{n}(-1)^p g_p = (-1)^n \chi(\mathcal{K}_{\widetilde{X}})\,,$$

where

$$g_p = h^{p,0} = \dim \Omega^p_{\widetilde{X}}(\widetilde{X})\,.$$

We obtain

4.7₄ Corollary. *The arithmetic genus is given by*

$$g = 1 + (-1)^n \dim[\Gamma(2,\ldots,2)]_0\,.$$

The Final Formulae. Let $a \in \mathbf{H}^n$ be an elliptic fixed point of Γ. We define the contribution $E(\Gamma, a)$ by

$$E(\Gamma, a) = \frac{1}{\#\Gamma_a} \sum_{\substack{M \in \Gamma_a \\ M \neq \text{id}}} N\frac{1}{1-\zeta}\,,$$

where $\zeta = (\zeta_1, \ldots, \zeta_n)$ are the rotation factors belonging to M ("$M \neq \text{id}$" means that the transformation induced by M is not the identity, i.e.

$$M \neq (\pm E, \ldots, \pm E)\,.)$$

If κ is a cusp, the contribution $L(\Gamma, \kappa)$ has been defined in §3 as a certain Shimizu L-series.

4.8 Theorem. *Let* $\Gamma \subset SL(2, \mathbf{R})^n$ *be a discrete subgroup such that the extended quotient* $(\mathbf{H}^n)^*/\Gamma$ *is compact. We assume that the first irreducibility* *assumption I.2.13 is satisfied, and, if H^n/Γ is compact, also the second irreducibility assumption I.4.12. Then the following formula holds*

$$1 + (-1)^n \dim[\Gamma, (2\ldots, 2)]_0 = (-1)^n \mathrm{vol}(H^n/\Gamma) + \sum_a E(\Gamma, a) + \sum_\kappa L(\Gamma, \kappa),$$

where a (resp. κ) runs over a complete set of representations of Γ-equivalence classes of elliptic fixed points (resp. cusps). This expression also equals the arithmetic genus g of a desingularization of X.

A Simple Special Case. Assume that $n > 1$ is an odd number and that Γ is without elliptic fixed points. Then the formula simplifies

$$g = 1 - \dim[\Gamma, (2\ldots, 2)]_0 = -\mathrm{vol}(H^n/\Gamma)$$

So the genus is a negative number. As a consequence we have for example that the field of automorphic functions is never a rational function field under this assumption, because the genus would otherwise be 1.

§5 Numerical Examples in Special Cases

The numerical evaluation of the invariants occuring in Shimizu's function $P_\Gamma(r)$ is in general very complicated. In some cases it can be calculated explicitly. We collect some well-known results, mostly without proofs. For more details we refer to Hirzebruch's paper [30].

The "main term" of Shimizu's function comes from the volume $\mathrm{vol}(\Gamma)$ of a fundamental domain of Γ with respect to the invariant measure

$$d\omega = \frac{1}{(4\pi)^n} \frac{dv}{Ny^2}$$

$$dv = \text{Euclidean measure} = dx_1\, dy_1 \ldots dx_n\, dy_n\,.$$

This volume has been calculated by Siegel [59] in the case of the Hilbert modular group. To be more precise, he expressed it in terms of the Dedekind ζ-function. Let K be a totally real number field of degree n. Its Dedekind ζ-function is defined as

$$\zeta_K(s) = \sum (\mathcal{N}(\mathbf{a}))^{-s}\,,$$

where the sum is extended over all integral ideals $\mathbf{a}$ $(0 \neq \mathbf{a} \subset \mathbf{o})$. This series converges if the real part of s is greater than 1 and defines an analytic function in this half-plane. It has an analytic continuation as a meromorphic function into the whole s-plane with a single pole (of first order) at $s = 1$.

The function

$$\xi_K(s) = d_K^{s/2}\pi^{-sn/2}\Gamma(s/2)^n\zeta_K(s) \qquad (d_K = \text{discriminant of } K)$$

is invariant under the transformation $s \to 1 - s$. Siegel's result is [59]:

5.1 Proposition. *The volume of a fundamental domain of Hilbert's modular group with respect to the invariant measure*

$$d\omega = (4\pi)^{-n}\frac{dv}{Ny^2} \qquad (dv = \text{Euclidean volume element})$$

is

$$\text{vol}(SL(2,\mathbf{o})) = 2^{1-n}(-1)^n\zeta_K(-1).$$

By means of the functional equation we obtain

$$(-1)^n\zeta_K(-1) = d_K^{3/2}2^{-n}\pi^{-2n}\zeta_K(2).$$

The trivial estimation

$$\zeta_K(2) > 1$$

gives us

5.2 Corollary.

$$\text{vol}(SL(2,\mathbf{o})) > 2^{1-2n}\pi^{-2n}d_K^{3/2}.$$

Explicit formulae for $\zeta_K(-1)$ are known in the case of a real quadratic field

$$K = \mathbf{Q}(\sqrt{a}),$$

where $a > 1$ is a square-free natural number. The discriminant of K is given by

$$d = 4a \quad \text{if } a \equiv 2,3 \mod 4$$
$$d = \ a \quad \text{if } a \equiv 1 \ \mod 4.$$

For the following result we refer to [30], p.192, where further references are given.

5.3 Proposition. *Let*

$$K = \mathbf{Q}(\sqrt{a}) \qquad (a > 1, \ square\text{-}free)$$

be a real quadratic field. Then for $a \equiv 1 \bmod 4$

$$\zeta_K(-1) = \frac{1}{30} \sum_{\substack{1 \le b < \sqrt{a} \\ b \text{ odd}}} \sigma_1\left(\frac{a - b^2}{4}\right)$$

and for $a \equiv 2, 3 \bmod 4$

$$\zeta_K(-1) = \frac{1}{60}\left(\sigma_1(a) + 2 \cdot \sum_{1 \le b < \sqrt{a}} \sigma_1(a - b^2)\right).$$

Here $\sigma_1(x)$ denotes the sum of the divisors of a natural number x.

The Elliptic Fixed Points. Following a method of Shimizu, Prestel succeeded in obtaining very explicit formulae for the elliptic fixed points of the Hilbert modular group $SL(2, \mathbf{o}_K)$ of a real quadratic field $K = \mathbf{Q}(\sqrt{a})$ (a square-free). In the following we describe some of these results, especially the cases a prime, $a > 5$. We first explain some notations.

Let $a \in \mathbf{H}^2$ be an elliptic fixed point of

$$\Gamma = SL(2, \mathbf{o}_K).$$

The order of a is the order of the group of mappings corresponding to Γ_a, i.e.

$$e := \operatorname{order}(a) = \#\Gamma_a/\{\pm E\}.$$

We denote by

$$a_e(\Gamma)$$

the number of Γ-equivalence classes of elliptic fixed points of order e. We recall the notion of the "type" of an elliptic fixed point: The stabilizer $\Gamma_a/\{\pm E\}$ is a cyclic group. After the transformation

$$z \to (z - a)(z - \bar{a})^{-1} = w$$

it is generated by

$$w \to \zeta w,$$

where

$$\zeta = (\zeta_1, \zeta_2)$$

is a pair of primitive roots of unity of order e. We especially have

$$\zeta_2 = \zeta_1^{e_2}; \quad 1 \le e_2 < e.$$

We call

$$(e_1, e_2) = (1, e_2)$$

the type of the elliptic fixed point.

Special Cases

1) $e = 2$: We have only one type, namely $(1,1)$.
2) $e = 3$: There are two types:
   Type I: $(1,1)$,
   Type II: $(1,2)$, .

In the following results of Prestel

$$h(-x) \qquad (x > 1)$$

denotes the class number of the imaginary quadratic field $\mathbf{Q}(\sqrt{-x})$.

5.4 Proposition. *Let $a > 1$ be a square-free natural number*

$$a > 5, \quad (a,6) = 1.$$

There exist only elliptic fixed points of order 2 and 3 for the Hilbert modular group $\Gamma = SL(2,\mathfrak{o})$ of the real quadratic field $K = \mathbf{Q}(\sqrt{a})$. Their number A_2, A_3 is given by the following formulae

1) $a \equiv 1 \bmod 4$: $\quad A_2 = h(-4a)$, $\quad A_3 = h(-3a)$.
2) $a \equiv 3 \bmod 8$: $\quad A_2 = 10h(-a)$, $\quad A_3 = h(-12a)$.
3) $a \equiv 7 \bmod 8$: $\quad A_2 = 4h(-a)$, $\quad A_3 = h(-12a)$.

In all three cases the fixed points of order 3 occur in pairs (one of type $(1,1)$ together with one of type $(1,2)$) *.

We next determine the contribution of the elliptic fixed points of order 2 and 3 to the rank formula

$$P_\Gamma(r) = \dim[\Gamma, (2r, \ldots, 2r)], \qquad r > 1.$$

We restrict to the case

$$r \equiv 0 \bmod 6,$$

because this is sufficient to determine the arithmetic genus (equivalently $\dim[\Gamma, (2, \ldots, 2)]$).

1) $e = \text{order}(a) = 2$

The contribution to the rank formula is

$$E(\Gamma, a) \quad (= E_2(\Gamma, a)) = \frac{1}{2} \frac{1}{(1+1)(1+1)} = \frac{1}{8}.$$

2) $e = \text{order}(a) = 3$

---

*   This is different if one also considers d divisible by 3 (see for example [30], p.237).

Type I $((e_1, e_2) = (1, 1))$:

Let ζ be a third root of unity. The contribution is

$$E(\Gamma, a) = \frac{1}{3} \left(\frac{1}{(1 - \zeta)^2} + \frac{1}{(1 - \zeta^2)^2} \right) .$$

Type II $((e_1, e_2) = (1, 2))$:

$$E(\Gamma, a) = \frac{1}{3} \left(\frac{1}{(1 - \zeta)(1 - \zeta^2)} + \frac{1}{(1 - \zeta^2)(1 - \zeta)} \right) .$$

5.5 Lemma. *Let $\Gamma \subset SL(2, \mathbf{R})^2$ be a discrete irreducible subgroup such that the extended quotient $(\mathbf{H}^2)^*/\Gamma$ is compact. The contribution of the fixed points of order 2 and 3 to the rank formula for $\dim[\Gamma, (2r, \ldots, 2r)]$, $r \equiv 0 \bmod 6$, is given by*

1) $e = \operatorname{order}(a) = 2$

$$E(\Gamma, a) = 1/8 .$$

2) $e = \operatorname{order}(a) = 3$

   Type I $((e_1, e_2) = (1, 1))$

$$E(\Gamma, a) = 1/9$$

   Type II $((e_1, e_2) = (1, 2))$

$$E(\Gamma, a) = 2/9$$

The Contribution of the Cusps. We recall that the cusp classes of the Hilbert modular group $\Gamma = SL(2, \mathbf{o})$ are in 1-1-correspondence with the ideal classes. If

$$\kappa = \frac{a}{c} ; \quad a, c \in \mathbf{o}$$

is a cusp, then

$$\mathbf{a} := (a, c)$$

represents the corresponding ideal class. To transform κ to ∞ we choose a matrix

$$A = \begin{pmatrix} a & b \\ c & d \end{pmatrix} ; \quad ad - bc = 1 ; \quad b, d \in \mathbf{a}^{-1}$$

which has the property

$$A\infty = \kappa .$$

A simple calculation shows that the conjugate group $A^{-1}\Gamma A$ equals

$$\left\{ \begin{pmatrix} \alpha & \beta \\ \gamma & \delta \end{pmatrix} ; \quad \alpha\delta - \beta\gamma = 1; \quad \alpha \in \mathbf{o}, \delta \in \mathbf{o}, \beta \in \mathbf{a}^{-2}, \gamma \in \mathbf{a}^2 \right\}.$$

The translation module of this group is $\mathbf{a}^{-2}$, and the group of multipliers is the group of unit squares:

$$\Lambda = \{\varepsilon^2, \ \varepsilon \text{ unit in } \mathbf{o}\}.$$

So in the case $n > 1$ the contribution of the cusps to the rank formula is given by

$$\frac{i^n}{(2\pi)^n}\sqrt{d(\mathbf{a}^{-2})} \sum_{a:\mathbf{a}^{-2}/\Lambda} \frac{\operatorname{sgn} Na}{|Na|}$$

where $d(\mathbf{a})$ denotes the discriminant of a given ideal $\mathbf{a}$ which, in the totally real case, is nothing else but the square of the volume $\operatorname{vol}(\mathbf{a})$ (see App. I for the definition of $d(\mathbf{a})$).

We recall that this expression does not change if one replaces $\mathbf{a}$ by an equivalent ideal. It vanishes, of course, if there exists a unit

$$\varepsilon \in \mathbf{o}^* \text{ with } N\varepsilon = -1.$$

We hence make the

Assumption. *Each unit ε in $\mathbf{o}$ has positive norm ($N\varepsilon = +1$).*

Under this assumption $\operatorname{sgn} Na$ merely depends upon the principal ideal (a) generated by a. We hence may define the character

$$\psi((a)) := \operatorname{sgn} Na$$

on the group of all principal ideals. It is 1 if the principal ideal is generated by a totally positive element.

Notation: $\mathcal{H}^\dagger$ *is the subgroup of the group of all principal ideals, which consists of ideals generated by a totally positive element. So our character is defined on the factor group*

$$\psi := \mathcal{H}/\mathcal{H}^\dagger \to \{1, -1\}.$$

We denote by $\mathcal{I}$ the group of all ideals of K. The elements of the factor group $\mathcal{I}/\mathcal{H}$ are the **ideal classes** and those of $\mathcal{I}/\mathcal{H}^\dagger$ the so-called **narrow ideal classes**. Both groups are finite. We have

$$\mathcal{H}/\mathcal{H}^\dagger \subset \mathcal{I}/\mathcal{H}^\dagger.$$

Because $I/\mathcal{H}^\dagger$ is a finite abelian group, we may extend ψ to a character

$$\chi : I/\mathcal{H}^\dagger \to S^1 = \{\zeta \in \mathbf{C}, |\zeta| = 1\}.$$

We have proved:

5.6 Lemma. *Assume that each unit has positive norm. Then there exists a character χ on the group I of all ideals depending only on the narrow ideal class and satisfying*

$$\chi((a)) = \operatorname{sgn} Na.$$

It is worthwhile asking whether χ can be taken to be real. This is, of course, the case if

$$I/\mathcal{H}^\dagger \simeq I/\mathcal{H} \times \mathcal{H}/\mathcal{H}^\dagger,$$

especially if the order of $I/\mathcal{H}$, i.e. the class number h of K, is odd.

Using a character χ described in Lemma 5.6 we may rewrite the contribution $L(\Gamma, \kappa)$ as follows:

We recall that Λ is the group of all squares of units. This group is a subgroup of the group of all units of index 2^n. This gives us

$$\sum_{a:a^{-2}/\Lambda} \frac{\operatorname{sgn} Na}{|Na|} = 2^n \sum_{a^{-2}/o^*} \frac{\operatorname{sgn} Na}{|Na|},$$

because all units have positive norm by assumption. If a runs over a complete system of representatives of a^{-2}/o^*, then $aa^2 = x$ runs over all integral ideals in the class of a^2. This gives

$$\sum_{a^{-2}/o^*} \frac{\operatorname{sgn} Na}{|Na|} = \mathcal{N}(a^2)\overline{\chi}(a^2) \sum_x \frac{\chi(x)}{\mathcal{N}(x)},$$

where x runs over all integral ideals in the ideal class of a^2. Here the infinite sum means the limit value of the L-series

$$\sum \frac{\chi(x)}{\mathcal{N}(x)^s}$$

which converges for $s > 1$ at $s = 1$. The expression

$$d(b)\mathcal{N}(b)^{-2}$$

does not depend on the ideal b and hence equals the discriminant d_K of the field K. We obtain for the contribution of our cusp

$$\frac{i^n}{\pi^n} d_K^{1/2} \overline{\chi}(a^2) \sum_x \frac{\chi(x)}{\mathcal{N}(x)}.$$

If we make the further assumption that χ is real, we have $\chi(\mathbf{a}^2) = \chi(\mathbf{a})^2 = 1$ and obtain for the sum of the contributions of all cusps

$$\frac{i^n}{\pi^n} d_K^{1/2} L(1,\chi) \, ,$$

where $L(1,\chi)$ denotes the limit value of the L-series

$$L(s,\chi) = \sum_{\mathbf{a} \subset \mathbf{o}} \chi(\mathbf{a}) \mathcal{N}(\mathbf{a})^{-s} \, ,$$

where $\mathbf{a}$ runs over all integral ideals.

5.7 Proposition. *Assume that the norm of each unit is positive. Assume furthermore that the character in 5.6 can be chosen to be real (for example if the class number is odd). Then the contribution of all cusps to the rank formula is*

$$\frac{i^n}{\pi^n} d_K^{1/2} L(1,\chi) \, .$$

It is a well-known fact that this expression is always unequal zero.

We finally consider a very interesting special case, namely the case of a real quadratic field

$$K = \mathbf{Q}(\sqrt{p}), \quad p \text{ prime} \, .$$

The following beautiful formulae can be found in Hirzebruch's paper [30], 3.10, where further comments and references are given: In $\mathbf{Q}(\sqrt{p})$ a unit of negative norm exists if and only if

$$p = 2 \quad \text{or} \quad p \equiv 1 \bmod 4 \, .$$

Hence precisely in this case the contribution of the cusps to the rank formula is 0. In the case $p \equiv 3 \bmod 4$ the squares of ideals generate a subgroup of index 2 in the narrow class group $\mathcal{I}/\mathcal{H}^\dagger$. This implies that there is exactly one real character χ with the properties in Lemma 5.6. By means of the decomposition law of the field $\mathbf{Q}(\sqrt{p})$ one obtains

$$L(s,\chi) = L_{-4}(s) L_{-p}(s) \, ,$$

where the L_a are the usual Dirichlet L-functions. The so-called class number formula gives

$$L_{-4}(1) = \frac{2\pi}{4} 4^{-1/2} h(-4)$$

$$L_{-p}(1) = \frac{2\pi}{2} p^{-1/2} h(-p)$$

where $h(-4)$ ($=$class number of $\mathbf{Q}(i)$) is one.

We now have the complete formula for the arithmetic genus

$$g = 1 + \dim[\Gamma, (2, \ldots, 2)]_0$$
$$= 1 + \dim[\Gamma, (2, \ldots, 2)] - h$$

in the case $K = \mathbf{Q}(\sqrt{p})$, p prime, $p > 5$. In the remaining cases $p = 2, 3, 5$ the determination of the elliptic fixed points is also continued in the mentioned paper of Prestel [52]. (Actually those three special cases have been treated by Gundlach in an earlier paper [21].)

5.8 Theorem. *Let p be a prime, $K = \mathbf{Q}(\sqrt{p})$. The arithmetic genus*

$$g = 1 + \dim[\Gamma, (2, \ldots, 2)]_0$$
$$= 1 + \dim[\Gamma, (2, \ldots, 2)] - h$$

of the Hilbert modular surface with respect to the Hilbert modular group $\Gamma = SL(2, \mathbf{o})$ is given by the formulae

$$
\begin{array}{ll}
g = 1 & \text{for } p = 2, 3, 5, \\[4pt]
g = \dfrac{1}{2}\zeta_K(-1) + \dfrac{h(-4p)}{8} + \dfrac{h(-3p)}{6} & \text{for } p \equiv 1 \bmod 4, \\[8pt]
g = \dfrac{1}{2}\zeta_K(-1) + \dfrac{3}{4}h(-p) + \dfrac{h(-12p)}{6} & \text{for } p \equiv 3 \bmod 8, \quad p > 5, \\[8pt]
g = \dfrac{1}{2}\zeta_K(-1) + \dfrac{h(-12p)}{6} & \text{for } p \equiv 7 \bmod 8.
\end{array}
$$

The Group Γ^-. We define

$$\Gamma^- := N\Gamma N^{-1}, \quad N_1 = \begin{pmatrix} 1 & 0 \\ 0 & 1 \end{pmatrix}, N_2 = \begin{pmatrix} 1 & 0 \\ 0 & -1 \end{pmatrix}.$$

The action of Γ^- on $\mathbf{H}^2$ is equivalent to the action of Γ on the product of an upper and a lower half-plane (by means of the usual formulae).

Consider an element

$$a \in K \text{ with } a^{(1)} > 0, \quad a^{(2)} < 0.$$

The matrix

$$A = \begin{pmatrix} \alpha & 0 \\ 0 & \alpha^{-1} \end{pmatrix} \in SL(2, \mathbf{R}) \times SL(2, \mathbf{R})$$
$$\alpha = \left(\sqrt{a^{(1)}}, -\sqrt{-a^{(2)}} \right)$$

has obviously the following property:

The groups

$$A\Gamma^- A^{-1} \quad \text{and} \quad \Gamma$$

are commensurable. If there is a unit a with the above property, both groups are equal.

The structure of the elliptic fixed points of Γ^- is dual with that of Γ.

The numbers of classes of fixed points of a certain order are equal, but the types are changed. In case of fixed points of order 3 precisely the two types ((1,1) and (1,2)) are interchanged. But in the cases which we considered the elliptic fixed points of order 3 occur in pairs (5.4).

So in the formulae for the arithmetic genus g of Γ and g^- of Γ^- only the contributions of the cusps may differ. It is obvious that they precisely differ by a sign. This gives us the following beautiful result of Hirzebruch. (For sake of completeness we also include the special cases $p = 2$ and 3).

5.9 Theorem. *Let p be a prime. The difference of the arithmetic genera*

$$g^- - g = \dim[\Gamma, (2, \ldots, 2)]_0 - \dim[\Gamma^-, (2, \ldots, 2)]_0$$
$$= \dim[\Gamma, (2, \ldots, 2)] - \dim[\Gamma^-, (2, \ldots, 2)]$$

equals

$$\begin{array}{ll} 0 & \text{if } p = 2 \text{ or } 3 \text{ or } p \equiv 1 \bmod 4, \\ h(-p) & \text{if } p > 3 \text{ and } p \equiv 3 \bmod 4. \end{array}$$

This result implies for example, that the field of modular functions with respect to Γ^- is not a rational function field if $p > 3$.

Final Remark. One might conjecture that the main term in the formula 4.8 for the arithmetic genus is the term which comes from the volume of the fundamental domain in the following sense. Let

$$\Gamma_m \subset SL(2, \mathbf{R})^n$$

($n = n(m)$ may vary with m) be a sequence of groups commensurable with a Hilbert modular group, such that for the different m, m' with $n(m) = n(m')$ the groups Γ_m and $\Gamma_{m'}$ are not conjugate in $SL(2, \mathbf{R})^n$. One then might conjecture

1) $$\mathrm{vol}(\mathbf{H}^n / \Gamma_m) \to \infty \quad \text{if } m \to \infty$$

2) $$\frac{g(\Gamma_m) - (-1)^n \mathrm{vol}(\mathbf{H}^n / \Gamma_m)}{\mathrm{vol}(\mathbf{H}^n / \Gamma_m)} \to 0 \quad \text{if } m \to \infty.$$

This conjecture would imply that only finitely many conjugacy classes of groups Γ with rational function field exist. Well-known estimates for the class number of an imaginary quadratic field and the estimate 5.2 show that this conjecture is true for the sequence of the usual Hilbert modular groups of $K = \mathcal{Q}(\sqrt{p}), p$ prime (s. 5.8 for the formula of the arithmetic genus). The conjecture is unsettled even if one restricts to a fixed $n > 2$ and to a usual Hilbert modular group.

For the case $n = 1$ see [60].

Chapter III. The Cohomology of the Hilbert Modular Group

§1 The Hodge Numbers of a Discrete Subgroup $\Gamma \subset SL(2,\mathbf{R})^n$ in the Cocompact Case

In this section we compute the Hodge numbers

$$h^{p,q} = \dim \mathcal{H}^{p,q}(\mathbf{H}^n)^\Gamma,$$

where $\Gamma \subset SL(2,\mathbf{R})^n$ is a discrete subgroup with compact quotient $\mathbf{H}^n/\Gamma$. In the case where Γ has no elliptic fixed points all those numbers can be expressed by means of a simple invariant, namely the volume of a fundamental domain with respect to the invariant measure.

The results of this section are due to Matsushima and Shimura.

We consider open domains

$$D_1, \ldots, D_n \subset \mathbf{C}$$

of the complex plane equipped with a Hermitean metric, i.e. a positive C^∞-function

$$h_i : D_i \longrightarrow \mathbf{R}_+ = \{x \in \mathbf{R} \mid x > 0\}.$$

We may consider the "product metric"

$$h = \begin{pmatrix} h_1 & & 0 \\ & \ddots & \\ 0 & & h_n \end{pmatrix}$$

on the domain

$$D = D_1 \times \ldots \times D_n.$$

Via the identification

$$\mathbf{C}^n \longleftrightarrow \mathbf{R}^{2n}$$
$$(z_1, \ldots, z_n) \longleftrightarrow (x_1, y_1, \ldots, x_n, y_n)$$

the associate Riemannian metric is given by

$$g = \begin{pmatrix} h_1 & & & & 0 \\ & h_1 & & & \\ & & \ddots & & \\ & & & h_n & \\ 0 & & & & h_n \end{pmatrix}.$$

(See App. III, Sects. IX–XI.) Such a metric has the Kähler property (A III, Sect. XII). We especially have the relation

$$\Delta = 2\square = 2\,\overline{\square}\ .$$

We make use of this relation to prove

1.1 Lemma. *If a, b are subsets of $\{1, \ldots, n\}$ and if f is a C^∞-function on our domain $D\ (= D_1 \times \ldots \times D_n)$, we have*

$$\Delta(f dz_a \wedge d\bar{z}_b) = g dz_a \wedge d\bar{z}_b$$

with a certain function g.

(Recall:

$$dz_a = dz_{a_1} \wedge \ldots \wedge dz_{a_p}\,,$$

where

$$a = \{a_1, \ldots, a_p\}\,,\ 1 \le a_1 < \ldots < a_p \le n$$

(similarly $d\bar{z}_b$).)

1.1₁ Corollary. *A differential form*

$$\omega = \sum f_{a,b} dz_a \wedge d\bar{z}_b$$

is harmonic if and only if all the components

$$f_{a,b} dz_a \wedge d\bar{z}_b$$

are harmonic.

Proof. It is easy to see that

$$*(dz_a \wedge d\bar{z}_b) \in C^\infty(D) dz_{\bar{b}} \wedge d\bar{z}_{\bar{a}}\,,$$

where $\bar{a}, \bar{b}$ denote the complements of a, b in $\{1, \ldots, n\}$. From the definition of $\square$, $\overline{\square}$ we now obtain

$$\overline{\square}\,(f dz_a \wedge d\bar{z}_b) = \sum_{\bar{b}} f_{\bar{b}} dz_a \wedge d\bar{z}_{\bar{b}}$$

and

$$\Box(f dz_a \wedge d\bar{z}_b) = \sum_{\bar{a}} h_{\bar{a}} dz_{\bar{a}} \wedge d\bar{z}_b .$$

This information together with $\Delta = 2\Box = 2\,\bar{\Box}$ is enough to verify 1.1. $\Box$

We consider the product $\mathbf{H}^n$ of n upper half-planes equipped with the Poincaré metric:

$$h(z) = \begin{pmatrix} y_1^{-2} & & 0 \\ & \ddots & \\ 0 & & y_n^{-2} \end{pmatrix} .$$

We have three types of motions:

1) The transformations $z \mapsto Mz$, $M \in SL(2,\mathbf{R})^n$.
2) The permutations of the variables.
3) If $a \subset \{1,\ldots,n\}$ is some subset, we may define a transformation

$$(z_1,\ldots,z_n) \longmapsto (w_1,\ldots,w_n) ,$$

where

$$w_i = -\bar{z}_i \quad \text{for } i \in a ,$$
$$w_i = z_i \quad \text{for } i \notin a .$$

It can be shown that the group of motions is generated by these special ones.

We consider a discrete subgroup

$$\Gamma \subset SL(2,\mathbf{R})^n .$$

In this section we make the assumptions:

1.2 Assumption. *1) The quotient $\mathbf{H}^n/\Gamma$ is compact.*
2) There exists a subgroup $\Gamma_0 \subset \Gamma$ of finite index without elliptic fixed points.

(Actually 2) is a consequence of 1) by a result of A. Borel.)

We are interested in the linear spaces

$$\mathcal{H}^{p,q}(\Gamma) = \{\omega \in M_\infty^{p,q}(\mathbf{H}^n)^\Gamma \mid \Delta\omega = 0\} ,$$

especially in their dimensions, the so called **Hodge numbers**

$$h^{p,q} = \dim \mathcal{H}^{p,q}(\Gamma) .$$

(We recall that $M_\infty^{p,q}(\mathbf{H}^n)$ denotes the spaces of alternating differential forms of type (p,q) with C^∞-coefficients, see A III.)

1.3 Proposition. *Under the above assumptions 1.2 we have*
1) $h^{p,q} = \dim \mathcal{H}^{p,q}(\Gamma) < \infty$.
2) *If ω is a Γ-invariant harmonic differential form, we have*

$$\partial \omega = \bar{\partial} \omega = \partial * \omega = \bar{\partial} * \omega = 0 \,,$$

and obviously the converse.

This follows from the fact that the quotient $\mathbf{H}^n/\Gamma_0$ carries a natural structure as compact Kählerian variety.

1) is a consequence of general finiteness properties of linear elliptic differential operators and
2) is a simple application of Stokes's theorem.

We omit the details because we shall obtain a different proof of 1.3 by means of the theory of automorphic forms.

We want to derive an explicit formula for the action of the star operator, and for this purpose we introduce a special basis of differential forms.

Notation: Put

$$\omega_i = \frac{dz_i \wedge d\bar{z}_i}{y_i^2} \,,$$

more generally

$$\omega_a = \omega_{a_1} \wedge \ldots \wedge \omega_{a_\alpha}$$

where

$$a = \{a_1, \ldots, a_\alpha\} \,, \ 1 \le a_1 < \ldots < a_\alpha \le n \,.$$

If a, b, c are three pairwise disjoint subsets of $\{1, \ldots, n\}$, we set

$$\Omega(a, b, c) = dz_a \wedge d\bar{z}_b \wedge \omega_c \,.$$

Obviously these elements form a basis of $M_\infty^{p,q}(D)$, where

$$p = \#a + \#c \,, \ q = \#b + \#c \,.$$

One advantage of the Ω's is that they are closed:

$$\partial \Omega(a, b, c) = \bar{\partial} \Omega(a, b, c) = 0 \,.$$

Another advantage is that the star operator can be easily applied to the Ω's.

1.4 Lemma. *Let*

$$\{1, \ldots, n\} = a \cup b \cup c \cup d$$

be a disjoint decomposition. We have

$$*\Omega(a, b, c) = C\Omega(a, b, d)$$

where $C = C(a, b, c, d)$ is a certain number.

(The constant C, which is not interesting for us, can be computed:

$$C = (-1)^{n(n-1)/2}(2i)^n 2^{-\alpha-\beta-2\delta}(-1)^{\beta+\delta+(\alpha+\beta)(\gamma+\delta)} ,$$

where the orders of $a, b, \ldots$ are denoted by the corresponding Greek letters.)

Proof. We need some information about the pairing $< , >$ which is used in the definition of the star operator. We have

$$< dx_i, dx_j > = < dy_i, dy_j > = \delta_{ij} h_i^{-1}$$
$$< dx_i, dy_j > = < dy_i, dx_j > = 0$$

hence

$$< dz_i, d\bar{z}_j > = < d\bar{z}_i, dz_j > = 2\delta_{ij} y_i^2$$
$$< dz_i, dz_j > = < d\bar{z}_i, d\bar{z}_j > = 0$$

In general

$$< \Omega(a, b, c), \Omega(\tilde{a}, \tilde{b}, \tilde{c}) >$$

is defined as the determinant of a certain $m \times m$-matrix $(m = \alpha + \beta + 2\gamma)$.

If $(\tilde{a}, \tilde{b}, \tilde{c}) \neq (b, a, c)$ this matrix contains less then m non-zero components. Its determinant therefore is zero. In the remaining case we obtain

$$y_a^2 \cdot y_b^2 \cdot \det \begin{pmatrix} 0 & 2E^{(\alpha)} & 0 & 0 \\ 2E^{(\beta)} & 0 & 0 & 0 \\ 0 & 0 & 0 & 2E^{(\gamma)} \\ 0 & 0 & 2E^{(\gamma)} & 0 \end{pmatrix} = (-1)^{\alpha\beta+\gamma} 2^{\alpha+\beta+2\gamma} y_a^2 \cdot y_b^2 .$$

The star operator is defined by the formula

$$< *\omega, \omega' > \omega_0 = \omega \wedge \omega' .$$

With the information we obtained about the pairing it is easy to verify

$$< C\Omega(a, b, d), \Omega(\tilde{a}, \tilde{b}, \tilde{d}) > \cdot \omega_0 = \Omega(a, b, c) \wedge \Omega(\tilde{a}, \tilde{b}, \tilde{d}) .$$

Both sides are zero except when

$$(\tilde{a}, \tilde{b}, \tilde{d}) = (b, a, d) ,$$

and in the latter case both sides are $(y_c y_d)^{-2}$ up to constant factors. □

We now obtain

1.5 Lemma. *Let ω be a differential form of the type*

$$\omega = f\Omega(a, b, c) .$$

We have

a) $\partial\omega = 0 \Leftrightarrow f$ is antiholomorphic in the variables $z_j, j \in b \cup d$.

b) $\bar{\partial}\omega = 0 \Leftrightarrow f$ is holomorphic in the variables $z_j, j \in a \cup d$.

c) $\partial(*\omega) = 0 \Leftrightarrow f$ is antiholomorphic in the variables $z_j, j \in b \cup c$.

d) $\bar{\partial}(*\omega) = 0 \Leftrightarrow f$ is holomorphic in the variables $z_j, j \in a \cup c$.

1.5₁ Corollary. *The relations*

$$\partial\omega = \bar{\partial}\omega = \partial * \omega = \bar{\partial} * \omega = 0$$

are equivalent with: ω is holomorphic in the variables coming from a, antiholomorphic in the variables coming from b and locally constant in the variables coming from $c \cup d$.

(A function $\varphi(z)$ of one complex variable is called antiholomorphic if $z \longmapsto \varphi(\bar{z})$ is holomorphic.)

Proof. We have

$$\partial\omega = \partial f \wedge \Omega(a, b, c) = 0$$

iff

$$\frac{\partial f}{\partial z_j} = 0 \text{ for } j \in b \cup d.$$

This means – by the Cauchy-Riemann equations – that f is antiholomorphic in the z_j's ($j \in b \cup d$). This proves a) and similarly b). For c), d) one has to use 1.4.   □

The corollary 1.5₁ implies a certain cancellation rule.

Cancellation Rule. *If $\omega = f\Omega(a, b, c)$ satisfies the equation*

$$\partial\omega = \bar{\partial}\omega = \partial * \omega = \bar{\partial} * \omega = 0,$$

then the same is true of

$$f\,dz_a \wedge d\bar{z}_b$$

instead of ω and conversely.

We now want to transform "antiholomorphic variables" into holomorphic ones. For this purpose we consider the diffeomorphism

$$\sigma_b : \mathbf{H}^n \longrightarrow \mathbf{H}^n$$

$$z \longmapsto w = \sigma_b(z),$$

where

$$w_j = \begin{cases} z_j & \text{for } j \notin b \\ -\bar{z}_j & \text{for } j \in b. \end{cases}$$

The function

$$z \longmapsto f(\sigma_b z)$$

is holomorphic in all variables if $f\Omega(a, b, c)$ satisfies the condition a)–d) in 1.5.

What does Γ-invariance of $\omega = f\Omega(a, b, c)$ mean for the transformed function

$$g(z) = f(\sigma_b z)?$$

To express this we introduce the notation

$$\begin{pmatrix} a & b \\ c & d \end{pmatrix}^- = \begin{pmatrix} a & -b \\ -c & d \end{pmatrix} = \begin{pmatrix} 1 & 0 \\ 0 & -1 \end{pmatrix} \begin{pmatrix} a & b \\ c & d \end{pmatrix} \begin{pmatrix} 1 & 0 \\ 0 & -1 \end{pmatrix}.$$

Obviously $M \mapsto M^-$ defines an automorphism of $SL(2, \mathbf{R})$. If

$$M = (M_1, \ldots, M_n) \in SL(2, \mathbf{R})^n,$$

we define

$$N = M^b \qquad \text{by}$$

$$N_j = \begin{cases} M_j & \text{if } j \notin b, \\ M_j^- & \text{if } j \in b. \end{cases}$$

The groups

$$\Gamma^b = \{M^b \mid M \in \Gamma\} \subset SL(2, \mathbf{R})^n$$

satisfy the same assumptions 1.2 as Γ (but they are different from Γ in general; the quotients H^n/Γ and H^n/Γ^b are topologically equivalent but they carry different "analytic structures").

The Γ-invariance of $\omega = f\Omega(a, b, c)$ means

$$f(Mz) = \prod_{j \in a}(c_j z_j + d_j)^2 \prod_{j \in b}(c_j \bar{z}_j + d_j)^2 f(z) \quad \text{for } M \in \Gamma,$$

because the forms

$$\omega_i = \frac{dz_i \wedge d\bar{z}_i}{y_i^2}$$

are invariant.

For the function

$$g(z) = f(\sigma_b z)$$

we obtain

$$g(Mz) = \prod_{j \in a \cup b} (c_j z_j + d_j)^2 g(z) \text{ for } M \in \Gamma^b.$$

(Use $\sigma_b^{-1}(M\sigma_b z) = M^b z$.)

Holomorphic functions g with this transformation property are special examples of automorphic forms as considered in Chap. I.

Assuming a certain condition of irreducibility (I.4.12) we were able to show (I.4.13) that these functions vanish unless

$$a \cup b = \emptyset \quad \text{or}$$
$$a \cup b = \{1, \ldots, n\}.$$

In the first case we have the constant functions, in the second case automorphic forms of weight $(2, \ldots, 2)$. We now obtain the complete picture of the Hodge spaces $\mathcal{H}^{p,q}(\Gamma)$. There are two possibilities for a non-vanishing harmonic form

$$\omega = f\Omega(a, b, c).$$

Case 1. $a \cup b = \emptyset$. In this case we have

$$\omega = \text{const} \cdot \omega_c.$$

These forms are actually invariant – not only with respect to our discrete subgroup – but with respect to the whole group $SL(2, \mathbf{R})^n$, and they are harmonic as follows from 1.4 (and $\partial\Omega(a, b, c) = \bar{\partial}\Omega(a, b, c) = 0$). We collect these "universally invariant" forms in the so-called **universal part of the Hodge spaces**

$$\mathcal{H}^{p,q}_{\text{univ}} = \begin{cases} \sum_{\#a=p} \mathbf{C}\omega_a & \text{if } p = q \text{ (and } 0 \le p \le n), \\ 0 & \text{elsewhere}. \end{cases}$$

The dimension of this space in the case $0 \le p = q \le n$ is $\binom{n}{p}$.

Case 2. $a \cup b = \{1, \ldots, n\}$.

We have

$$\omega = f dz_a \wedge d\bar{z}_b,$$

and

$$g(z) = f(\sigma_b z)$$

is an automorphic form of weight $(2, \ldots, 2)$ with respect to Γ^b.

1.6 Theorem. *Let $\Gamma \subset SL(2, \mathbf{R})^n$ be a discrete subgroup which satisfies the assumption 1.2 (especially, that $\mathbf{H}^n/\Gamma$ is compact) and the irreducibility condition I.4.12. We have*

1) in the case $p + q \neq n$

$$\mathcal{H}^{p,q}(\Gamma) = \mathcal{H}^{p,q}_{\text{univ}}.$$

2) in the case $p + q = n$

$$\mathcal{H}^{p,q}(\Gamma) \cong \mathcal{H}^{p,q}_{\text{univ}} \bigoplus_{\substack{b \subset \{1,\ldots,n\} \\ \#b=q}} [\Gamma^b, (2,\ldots,2)] .$$

The dimensions of the spaces $[\Gamma^b, (2,\ldots,2)]$ have been computed in Chap. II by means of the Selberg trace formula in connection with an algebraic geometric method (to come down to the "border weight" $(2,\ldots,2)$).

1.6₁ Corollary. *If (in addition) Γ has no elliptic fixed points, the dimensions of the spaces $[\Gamma^b, (2,\ldots,2)]$ do not depend on b. We have*

$$\dim[\Gamma^b, (2,\ldots,2)] = \text{vol}(\mathbf{H}^n/\Gamma) + (-1)^{n+1} ,$$

where the volume is taken with respect to the invariant volume element.

As a special case of 1.6 we obtain the spaces

$$\mathcal{H}^m(\Gamma) = \{\omega \in M^m_\infty(D) \mid \Delta\omega = 0\}$$

of all harmonic Γ-invariant differential forms of degree m.

Notice. These spaces do not depend on the holomorphic structure but only on the underlying Riemannian metric.

From the equation $\Delta = 2\square$ we know that Δ is compatible with the (p,q)-bigraduation, hence

$$\mathcal{H}^m(\Gamma) = \bigoplus_{p+q=m} \mathcal{H}^{p,q}(\Gamma) .$$

The dimensions of these spaces are denoted by

$$b^m = \dim \mathcal{H}^m(\Gamma)$$
$$= \sum_{p+q=m} h^{p,q} .$$

They are 0 if $m > 2n$ (or $m < 0$).

1.7 Theorem. *Under the assumptions of 1.6 and its corollary we have*
1) in the case $m \neq n$

$$b^m = \begin{cases} \binom{n}{m/2} & \text{if } m \text{ is even,} \\ 0 & \text{if } m \text{ is odd.} \end{cases}$$

2) in the case $m = n$

$$b^m = 2^n \cdot \dim[\Gamma, (2,\ldots,2)] + \begin{cases} \binom{n}{n/2} & \text{if } n \text{ is even,} \\ 0 & \text{if } n \text{ is odd.} \end{cases}$$

1.7₁ Corollary. *The alternating sum of all the b^m is*

$$\sum_{j=0}^{2n}(-1)^j b^j = (-2)^n \cdot \text{vol}(\mathbf{H}^n/\Gamma)\,.$$

Final Remark. The numbers calculated above are actually dimensions of cohomology groups. From the general Hodge theory (App. III) follows

$$\mathcal{H}^m(\Gamma) \cong H^m(\mathbf{H}^n/\Gamma, \mathbf{C})$$

(singular cohomology with coefficients $\mathbf{C}$). It should be mentioned that the last formula (corollary of 1.7) is also a consequence of the Gauß-Bonnet formula which expresses the Euler characteristic (= alternating sum of Betti numbers) by means of the curvature and the volume. If Γ has no elliptic fixed points, one furthermore has

$$\mathcal{H}^{p,q}(\Gamma) \cong H^q(\mathbf{H}^n/\Gamma, \Omega^p)$$

where Ω^p denotes the sheaf of holomorphic p-forms on the analytic manifold $\mathbf{H}^n/\Gamma$.

§2   The Cohomology Group of the Stabilizer of a Cusp

Let

$$D \subset \mathbf{R}^n$$

be an open domain and Γ a group of C^∞-diffeomorphisms of D onto itself. We assume that Γ acts discontinuously and that Γ has a subgroup Γ_0 of finite index which acts freely on D. We denote by

$$M_\infty^p(D)^\Gamma$$

the linear space of all Γ-invariant C^∞-differential forms of degree p on D. We may consider the so-called "de Rham complex"

$$\ldots \longrightarrow M_\infty^p(D)^\Gamma \xrightarrow{\ d\ } M_\infty^{p+1}(D)^\Gamma \longrightarrow \ldots$$

("complex" means $d \cdot d = 0$) and the de Rham cohomology groups (they are actually $\mathbf{C}$-vector spaces)

$$H^p(\Gamma) = H^p((D, \Gamma)) = C^p/B^p\,,$$

where
$$C^p = \ker(M^p_\infty(D)^\Gamma \longrightarrow M^{p+1}_\infty(D)^\Gamma)$$
$$B^p = \operatorname{im}(M^{p-1}_\infty(D)^\Gamma \longrightarrow M^p_\infty(D)^\Gamma).$$

Of course we have

$$H^p((D,\Gamma)) = 0 \quad \text{if} \quad p < 0 \text{ or } p > n.$$

Notice. By the theorem of de Rham there is a natural isomorphism between $H^p((D,\Gamma))$ and the singular cohomology group $H^p(D/\Gamma, \mathbf{C})$. If D is contractible (for example if D is convex) we have furthermore

$$H^p((D,\Gamma)) \cong H^p(D/\Gamma, \mathbf{C})$$
$$\cong H^p(\Gamma, \mathbf{C})$$

where $H^\bullet(\Gamma, \mathbf{C})$ denotes the group cohomology of Γ acting trivially on $\mathbf{C}$.

We now assume that a discrete subgroup $\Gamma \subset SL(2,\mathbf{R})^n$ with cusp ∞ is given. We want to compute the cohomology of the stabilizer Γ_∞. Recall that the stabilizer Γ_∞ consists of transformations of the form

$$z \longmapsto \varepsilon z + b, \quad N\varepsilon = 1.$$

We have two types of differential forms which are closed and invariant under all thEse transformations, namely

1) $dx_1 \wedge \ldots \wedge dx_n,$

2) $\frac{dy_a}{y_a} := \frac{dy_{a_1}}{y_{a_1}} \wedge \ldots \wedge \frac{dy_{a_p}}{y_{a_p}},$
where
$$a = \{a_1, \ldots, a_p\}, \quad 1 \le a_1 < \ldots < a_p \le n.$$

The classes of the dy_a/y_a are not independent. The function $\sum_{j=1}^n \log y_j$ is invariant. Therefore

$$\frac{dy_n}{y_n} \quad \text{and} \quad -\left(\frac{dy_1}{y_1} + \ldots + \frac{dy_{n-1}}{y_{n-1}}\right)$$

define the same class in the de Rham cohomology group.

2.1 Proposition. *We obtain a basis of $H^m(\Gamma_\infty)$ by the classes of the following differential forms*

a) $m < n$

$$(dy_a)/y_a, \quad a \subset \{1, \ldots, n-1\}, \quad \#a = m.$$

b) $m \ge n$

$$(dy_a)/y_a \wedge dx_1 \wedge \ldots \wedge dx_n, \quad a \subset \{1, \ldots, n-1\}, \quad \#a = m - n.$$

For the dimension b_m (Betti numbers) of the spaces we obtain

$$b_m = \binom{n-1}{m} \quad \text{if} \quad m < n,$$

$$b_m = \binom{n-1}{2n-1-m} \quad \text{if} \quad m \geq n.$$

There are several proofs of this proposition. Perhaps the most natural one is to consider the exact sequence

$$0 \longrightarrow \mathfrak{t} \longrightarrow \Gamma_\infty \longrightarrow \Lambda \longrightarrow 0$$

and the Hochschild-Serre spectral sequence or equivalently the spectral sequence corresponding to the fibration

$$H^n/\mathfrak{t} \longrightarrow H^n/\Gamma_\infty.$$

We hope that the reader will appreciate the following proof of 2.1 as a funny one:

Proof of 2.1. The group of all transformations

$$z \longrightarrow \varepsilon z + b, \quad N\varepsilon = 1,$$

also operates on

$$D = \{z \in H^n \mid Ny = 1\}.$$

We may identify D with $\mathbf{R}^{2n-1}$ by means of the mapping

$$D \overset{\sim}{\longrightarrow} \mathbf{R}^{2n-1}$$

$$z \longmapsto (x_1, \ldots, x_n, \underbrace{\log y_1}_{=u_1}, \ldots, \underbrace{\log y_{n-1}}_{=u_{n-1}}).$$

We especially may consider the de Rham cohomology of (D, Γ_∞). Pulling back differential forms by means of the natural inclusion

$$\mathbf{R}^{2n-1} \cong D \hookrightarrow H^n$$

gives a mapping

$$H^p((H^n, \Gamma_\infty)) \longrightarrow H^p((D, \Gamma_\infty)).$$

We make use of the fact that this is an isomorphism for all p.

The spaces H^n/Γ_∞ and $\mathbf{R} \times D/\Gamma_\infty$ are obviously homeomorphic. The natural imbedding

$$D/\Gamma_\infty \hookrightarrow H^n/\Gamma_\infty$$

is therefore a homotopy equivalence and induces isomorphisms in singular cohomology. The statement now follows from the de Rham theorem.

It is possible to "imitate" the homotopy argument directly in the de Rham complex by integrating along $t = \sqrt[p]{y_1 \cdots \cdot y_n}$.

The formula

$$dg(x) = f(x)dx\,,$$

where

$$g(x) = \int_0^x f(t)dt$$

shows for example that the first de Rham cohomology group of **R** vanishes.

The quotient D/Γ_∞ is compact! We therefore can use Hodge theory to compute the de Rham cohomology. For this purpose we choose any Riemannian metric g on D ($\cong \mathbf{R}^{2n-1}$) which is invariant under all transformations

$$z \longrightarrow \varepsilon z + b\,, \quad N\varepsilon = 1$$

(not only under the given discrete subgroup Γ_∞). Such a metric exists (for example the restriction of the Poincaré metric on $\mathbf{H}^n$).

We now consider a whole sequence of groups, namely

$$G(l) = \begin{pmatrix} l^{-1} & 0 \\ 0 & l \end{pmatrix} \Gamma_\infty \begin{pmatrix} l & 0 \\ 0 & l^{-1} \end{pmatrix}\,, \quad l = 1, 2, 3, \ldots$$

The de Rham cohomology groups of Γ_∞ and $G(l)$ are of course isomorphic. Before we continue we make the following assumption

Assumption. The group Γ_∞ splits, i.e. if ε is a multiplier, then the transformation $z \longrightarrow \varepsilon z$ is contained in Γ_∞ (and not only some transformation of the form $z \longmapsto \varepsilon z + b_\varepsilon$ as demanded in the definition of "cusp ∞"). We shall show at the end of the proof how the general case can be reduced to the split case.

In the split case $G(l)$ consists of all transformation

$$z \longmapsto \varepsilon z + a/l^2\,,$$
$$\varepsilon \in \Lambda \quad \text{(group of multipliers of } \Gamma_\infty)$$
$$a \in \mathfrak{t} \quad \text{(group of translations of } \Gamma_\infty)\,.$$

They contain all the transformations of $G(1) = \Gamma_\infty$, because $l^2\mathfrak{t} \subset \mathfrak{t}$. If we denote by $\mathcal{H}^m(G(l))$ the space of all harmonic m-forms on D which are $G(l)$-invariant, we obtain

$$\mathcal{H}^m(G(l)) \subset \mathcal{H}^m(\Gamma_\infty)\,.$$

Equality holds, because the dimensions are the same. The union of all $l^{-2}\mathfrak{t}$ is dense in $\mathbf{R}^n$. We hence obtain by a continuity argument that if

$$\omega = \sum f_{a,b}(x, u)du_a \wedge dx_b$$

*is a Γ_∞-invariant harmonic differential form, then the functions $f_{a,b}(x,u)$
do not depend on x.* The invariance under the multipliers

$$z \longmapsto \varepsilon z$$

now gives us

$$f_{a,b} \neq 0 \Longrightarrow b = \emptyset \text{ or } b = \{1,\ldots,n\} \,.$$

We treat the first case $b = \emptyset$ (the case $b = \{1,\ldots,n\}$ is similar):

We now have a closed differential form of the type

$$\omega = \sum_{a \subset \{1,\ldots,n-1\}} f_a(u) du_a \quad (u = (u_1,\ldots,u_{n-1}))$$

and want to show that these forms are cohomologeous to a linear combination

$$\sum C_a(dy_a)/y_a \,.$$

This means that the difference is of the form $d\tilde\omega$, where $\tilde\omega$ denotes a Γ_∞-invariant differential form on D. It is easily proved that no x-variables occur in $\tilde\omega$. Hence we are faced with the problem of determining the de Rham cohomology of Λ acting on $\mathbf{R}_+^{n-1}$ or, equivalently, of the lattice $\log \Lambda$ acting by translations on $\mathbf{R}^{n-1}$.

The cohomology of a lattice $L \subset \mathbf{R}^m$ (acting by translation) can be determined by means of the same trick as above. Replacing L by $(1/l)^2 L$ and using the usual Euclidean metric ($g = E = $ unit matrix) one shows that each harmonic form which is periodic under L is necessarily of the form

$$\omega = \sum_{a \subset \{1,\ldots,m\}} C_a du_a$$

with constants C_a. It is easy to see that the classes of du_a are linearly independent. If we transform du_a by means of

$$\log : \mathbf{R}_+^n \longrightarrow \mathbf{R}^n$$
$$y \longmapsto u = \log y \,,$$

we obtain the differential forms $(dy_a)/y_a$ instead of du_a. $\square$

We finally show how to avoid the assumption that Γ_∞ splits.

First of all we may assume that at least for one multiplier $\varepsilon \in \Lambda, \varepsilon \neq 1$, the transformation

$$z \longmapsto \varepsilon z$$

is contained in Γ_∞, because we may conjugate Γ_∞ by means of a translation

$$z \longmapsto z + \alpha \,,$$

where

$$\alpha = b_\varepsilon (1 - \varepsilon)^{-1} .$$

We now claim

2.2 Remark. *If Γ_∞ contains at least one transformation of the form*

$$z \longmapsto \varepsilon_0 z , \qquad \varepsilon_0 \neq 1 ,$$

then there exists a natural number d such that each transformation of Γ_∞ is of the form

$$z \longmapsto \varepsilon z + b ,$$

$$\varepsilon \in \Lambda \quad \text{(group of multipliers of } \Gamma_\infty)$$
$$\text{where } db \in \mathfrak{t} \quad \text{(group of translations of } \Gamma_\infty).$$

Proof. Let

$$z \longmapsto \varepsilon z + b$$

be any transformation of Γ_∞. We see that

$$z \longmapsto \varepsilon \varepsilon_0 z + b$$

and

$$z \longmapsto (\varepsilon z + b) \cdot \varepsilon_0$$

are both contained in Γ_∞. Multiplying one with the inverse of the other we obtain

$$b(1 - \varepsilon_0) \in \mathfrak{t} .$$

We know that Λ acts by multiplication on $\mathfrak{t}$, especially

$$(1 - \varepsilon_0)\mathfrak{t} \subset \mathfrak{t} .$$

The index has to be finite, because both groups are isomorphic to $\mathbf{Z}^n$. Therefore there exists a natural number d such that

$$d\mathfrak{t} \subset (1 - \varepsilon_0)\mathfrak{t}$$

and this implies

$$b \in d^{-1}\mathfrak{t} . \qquad \qquad \square$$

Remark 2.2 shows us that the groups

$$G(l) = \begin{pmatrix} l^{-1} & 0 \\ 0 & l \end{pmatrix} \Gamma_\infty \begin{pmatrix} l & 0 \\ 0 & l^{-1} \end{pmatrix}$$

contain Γ_∞ if l is divisible by d. This is enough to carry through the argument used during the proof in the split case.

§3 Eisenstein Cohomology

Let $\Gamma \subset SL(2,\mathbf{R})^n$ be a discrete subgroup such that $(\mathbf{H}^n)^*/\Gamma$ is compact. We investigate the natural mapping

$$H^m(\Gamma) \longrightarrow \overset{h}{\underset{j=1}{\bigoplus}} H^m(\Gamma_{\kappa_j}),$$

where $\kappa_1,\ldots,\kappa_h$ denotes a system of representatives of the cusps. In the case of a congruence group of a Hilbert modular group we will construct a certain subspace of $H^m(\Gamma)$ which maps isomorphically to the image of $H^m(\Gamma)$ by means of Eisenstein series. This theory is due to Harder [26].

We start with the basis of $H^m(\Gamma_\infty)$ which has been constructed in §2. Let ω be one of the basis elements, i.e.

1) $$\omega = \frac{dy_a}{y_a}, \quad a \subset \{1,\ldots,n-1\}, \quad (m < n),$$

2) $$\omega = \frac{dy_a}{y_a} \wedge dx_1 \wedge \ldots \wedge dx_n, \quad a \subset \{1,\ldots,n-1\}, \quad (m \geq n).$$

It is convenient to replace the differential forms in case 2) by a basis which fits better into our setting of Eisenstein series.

3.1 Remark. *Assume $n \leq m < 2n$. The differential forms*

$$\omega = \frac{dz_a \wedge d\overline{z}_a}{y_a} \wedge dz_b,$$

where

$$a \cup b = \{1,\ldots,n\}, \quad a \cap b = \emptyset, \quad \#a = m-n,$$

are Γ_∞-invariant and closed. Their classes define a basis of $H^m(\Gamma_\infty)$.

Proof. We show that the differential forms

$$\frac{dz_a \wedge d\overline{z}_a}{y_a} \wedge dz_b$$

and

$$\frac{dy_a}{y_a} \wedge dx_1 \wedge \ldots \wedge dx_n$$

are cohomologous upto a constant factor. This obviously follows from the fact that a differential form of the type

$$\frac{dy_a}{y_a} \wedge dx_b \wedge dy_c \, , \quad a \neq \{1,\ldots,n\} \, , \quad b \cup c = \{1,\ldots,n\} \, ,$$

defines the zero class if $c \neq \emptyset$.

This is trivial if $c \cap a \neq \emptyset$. But if $c \cap a = \emptyset$, we choose some index $i \in c$. The differential form

$$y_i \cdot \frac{dy_a}{y_a} \wedge dx_b \wedge dy_{c'} \, , \quad \text{where } c' = c - \{i\} \, ,$$

is Γ_∞-invariant, and its exterior derivative is up to a sign

$$\frac{dy_a}{y_a} \wedge dx_b \wedge dy_c \, . \qquad \qquad \square$$

Let now ω be one of our basis elements, i.e.

Case 1: $\omega = (dy_a)/y_a$,

Case 2: $\omega = \frac{dz_a \wedge d\bar{z}_a}{y_a} \wedge dz_b \quad (a \cup b = \{1,\ldots,n\} \, , \ a \subset \{1,\ldots,n-1\})$.

These forms are Γ_∞-invariant, but we want to have Γ-invariant forms. It looks natural to construct them by symmetrization:

$$E(\omega) := \sum_{M \in \Gamma_\infty \backslash \Gamma} \omega \mid M \, .$$

We use the notation

$$\omega \mid M := M^* \omega \, .$$

We have

$$(\omega \mid M) \mid N = \omega \mid MN \, ,$$

and hence $\omega \mid M$ does not depend on the choice of the representative M. The main problem will be the question of convergence. The formulae

$$dz \mid M \ = \ (cz + d)^{-2} dz \, ,$$

$$d\bar{z} \mid M \ = \ (c\bar{z} + d)^{-2} d\bar{z} \, ,$$

$$y \mid M \ = \mid cz + d \mid^{-2} \cdot y \, ,$$

$$dy \ = \ (1/2i)(dz - d\bar{z})$$

show that the series $E(\omega)$ can be expressed by Eisenstein series of the following type:

We consider two vectors $\alpha, \beta \in \mathbf{Z}^n$ of integers. We assume that

$$2r = \alpha_j + \beta_j \quad (1 \le j \le n)$$

is independent of j and that r is integral. We then may (formally) consider the Eisenstein series

$$E_{\alpha,\beta}(z) = E_{\alpha,\beta}^{\Gamma}(z) = \sum_{M \in \Gamma_\infty \backslash \Gamma} N(cz+d)^{-\alpha} N(c\bar{z}+d)^{-\beta}.$$

In Chap. I, §5 we already considered the case $\beta = 0$. We now obtain (formally)

$$E(\omega) = \varphi(z) \cdot \omega,$$

where $\varphi(z)$ is a linear combination of the Eisenstein series introduced before. More precisely

$$\varphi(z) = \sum C_{\alpha,\beta} E_{\alpha,\beta}(z).$$

For the total weights

$$2r = \alpha_j + \beta_j$$

we obtain:

Case 1: $r = 0$ $(\omega = \sum C_a \frac{dy_a}{y_a})$,

Case 2: $r = 1$ $(\omega = \sum C_a \frac{dz_a \wedge d\bar{z}_a}{y_a} \wedge dz_b)$.

From our assumption $a \subset \{1, \ldots, n-1\}$ we furthermore obtain in the second case

$$\alpha \ne \beta.$$

Up to now our consideration has been formal. We now have to deal with the question of convergence. We have already proved that

$$\sum | N(cz+d) |^{-2r}$$

converges for all real $r > 1$. The same proof shows that this series does **not converge** if $r = 1$.

In the first case ($2r = \alpha_j + \beta_j = 0$) we are rather away from the border of convergence.

The second case looks better. Here we are precisely at the border of absolute convergence ($r = 1$). Following an idea of Hecke we can define $E_{\alpha,\beta}(z)$ in this case as follows: We first introduce for real $s > 0$ the series

$$E_{\alpha,\beta}(z,s) := \sum_{\Gamma_\infty \backslash \Gamma} N(cz+d)^{-\alpha} N(c\bar{z}+d)^{-\beta} | N(cz+d) |^{-2s}.$$

This series converges absolutely (in the case $2r = \alpha_j + \beta_j = 2, s > 0$).

One may ask whether the limit

$$\lim_{s \to 0} E_{\alpha,\beta}(z,s) \quad (s > 0)$$

exists. If this happens we say:

The series $E_{\alpha,\beta}(z)$ admits Hecke summation. The value of this series is defined by

$$E_{\alpha,\beta}(z) = \lim_{s \to 0} E_{\alpha,\beta}(z,s) \quad (s > 0).$$

We keep the notation

$$E_{\alpha,\beta}(z) \; " = " \; \sum N(cz+d)^{-\alpha} N(c\bar{z}+d)^{-\beta},$$

where the symbol "=" indicates that we applied Hecke summation.

In the next section (§4: Analytic continuation of Eisenstein series) we deal with the question of Hecke summation. We shall show that in the case of a congruence subgroup Γ (i.e., a discrete subgroup of $SL(2,\mathbf{R})^n$ which contains a principal congruence subgroup $\Gamma_K[\mathfrak{a}]$ of some Hilbert modular group Γ_K. A deep Theorem by A. Selberg states that each discrete subgroup with a fundamental domain of finite volume and with at least one cusp is conjugate to a congruence subgroup.) Hecke summation always exists. We hence assume in the following that Γ is a congruence subgroup. *Let*

$$\alpha, \beta \in \mathbf{Z}^n$$

be two vectors of integers with the properties

a)
$$\alpha \neq \beta,$$

b)
$$\alpha_j + \beta_j = 2 \quad for \quad j = 1, \ldots, n.$$

Then the Eisenstein series

$$E_{\alpha,\beta}(z) \; " = " \; \sum N(cz+d)^{-\alpha} N(c\bar{z}+d)^{-\beta}$$

can be defined by Hecke summation. We now return to the symmetrization of the differential form

$$\omega = \frac{dz_a \wedge d\bar{z}_a}{y_a} \wedge dz_b.$$

We have

$$\omega \mid M = \omega \cdot N(cz+d)^{-\alpha} N(c\bar{z}+d)^{-\beta},$$

where

$$\alpha_j = \begin{cases} 1 & \text{if } j \in a \\ 2 & \text{if } j \in b, \end{cases}$$

$$\beta_j = \begin{cases} 1 & \text{if } j \in a \\ 0 & \text{if } j \in b. \end{cases}$$

The results about Eisenstein series described above show that the differential form

$$E(\omega) \text{ " } = \text{ " } \sum_{M \in \Gamma_\infty \backslash \Gamma} \omega \mid M \text{ " } = \text{ " } \omega \cdot \sum N(cz + d)^{-\alpha} N(c\bar{z} + d)^{-\beta}$$

can be defined by Hecke summation.

3.2 Proposition. *Let ω be one of the basis elements described in 3.1. Then the differential form*

$$E(\omega) \text{ " } = \text{ " } \sum_{M \in \Gamma_\infty \backslash \Gamma} \omega \mid M := \lim_{s \to 0} \sum_{M \in \Gamma_\infty \backslash \Gamma} \mid N(cz + d) \mid^{-2s} \omega \mid M \quad (s > 0)$$

exists.

We now have to investigate whether the differential form form $E(\omega)$ is closed or not. A differential form

$$f(z) \frac{dz_a \wedge d\bar{z}_a}{y_a} \wedge dz_b \quad (a \cup b \subset \{1, \dots, n\})$$

is obviously closed if and only if $f(z)$ is holomorphic in the variables coming from b.

Recall:

$$j \in b \quad \Longleftrightarrow \quad \beta_j = 0.$$

In the next section (§4, Theorem 4.9) we prove the existence of a real number B such that

$$E_{\alpha,\beta}(z) - B/Ny$$

is holomorphic in all variables z_j with $\beta_j = 0$.

As a consequence, the differential form $E(\omega)$ is closed if and only if the number B is 0. We also will obtain some information about the constant B: It is 0 if the number of all j such that

$$\alpha_j = \beta_j = 1$$

is less or equal $n - 2$. In our situation we have

$$\alpha_j = \beta_j = 1 \quad \Longleftrightarrow \quad j \in a$$

and

$$m = \#a + n = 2n - \#b.$$

This means that B is zero if

$$\#b \geq 2 \quad \text{or} \quad m \leq 2n - 2.$$

3.3 Proposition. *Assume*

$$n \leq m \leq 2n - 2.$$

The differential form

$$E(\omega) \text{ "} = \text{"} \sum_{M \in \Gamma_\infty \backslash \Gamma} \omega \mid M,$$

where

$$\omega = \sum C_a \frac{dz_a \wedge d\bar{z}_a}{y_a} \wedge dz_b, \quad a \cup b = \{1, \ldots, n\}, \quad m = \#a + n$$

(which can be constructed by means of Hecke summation of Eisenstein series) has the following properties:
1) $E(\omega)$ is closed.
2) $E(\omega)$ and ω define the same cohomology class in $H^m(\Gamma_\infty)$.
3) The image of $E(\omega)$ in $H^n(\Gamma_\kappa)$ is 0 if κ is a cusp inequivalent to ∞.

Proof. We know already that $E(\omega)$ is closed, because it depends holomorphically on the variables coming from b. It is especially a holomorphic function of z_n. The Fourier expansion of $E(\omega) \mid M$ (4.9) may be rewritten in the form

$$E(\omega) \mid M - A\omega = \omega \cdot \sum_{g \in \mathfrak{t}^\circ, g_n > 0} b_g(z_1, \ldots, z_{n-1}) e^{2\pi i g_n z_n},$$

where the coefficients b_g are independent of z_n. Integration of the series as a function of z_n term by term gives us a function

$$f(z) = \sum_{g \in \mathfrak{t}^\circ, g_n > 0} b_g(z_1, \ldots, z_{n-1}) / 2\pi i g_n \cdot e^{2\pi i g_n z_n}.$$

We now consider the differential form

$$\omega' := f(z) \frac{dz_a \wedge d\bar{z}_a}{y_a} \wedge dz_{b'},$$

where $b' = b - \{n\}$.

3.3₁ Remark. *The differential form ω' is invariant under $(M\Gamma M^{-1})_\infty$.*

The proof is easy, because invariance under transformations

$$z \longmapsto \varepsilon z + b$$

can be expressed by properties of the Fourier coefficients. The invariance of ω' will follow from that of $E(\omega) \mid M$. We leave the details to the reader.

We now obtain that

$$E(\omega) \mid M - A\omega$$

defines the zero class in

$$H^m((M\Gamma M^{-1})_\infty).$$

But from 4.9 we know that $A = 1$ if M is the unit matrix (hence $E(\omega) \sim \omega$ in $H^m(\Gamma_\infty)$) and $A = 0$ if the cusp $\kappa = M^{-1}(\infty)$ is inequivalent to ∞ (and hence $E(\omega) \sim 0$ in $H^m(\Gamma_\kappa)$). This completes the proof of 3.3.   □

By the method of transforming an arbitrary cusp to infinity one can easily generalize the results of Proposition 3.3 to arbitrary cusps:

Let

$$\kappa = M^{-1}\infty, \quad M \in SL(2,K),$$

be an arbitrary cusp and ω one of the basis elements described in 3.1. For the moment it is not necessary to assume $m < 2n - 1$.

We may consider Eisenstein series with respect to the conjugate group

$$\Gamma_M = M\Gamma M^{-1},$$

especially

$$E^{\Gamma_M}(\omega) \,\text{“}=\text{”}\, \sum_{N:(\Gamma_M)_\infty \backslash \Gamma_M} \omega \mid N.$$

The form

$$E^{\Gamma_M}(\omega) \mid M$$

is again Γ-invariant.

The one-dimensional space

$$(E^{\Gamma_M}(\omega) \mid M) \cdot \mathbf{C}$$

depends only on the Γ-equivalence class of κ.

3.4 Definition. *The space*

$$\mathcal{E}(\Gamma, m) \quad (m \geq n)$$

is generated by all differential forms

$$E^{\Gamma_M}(\omega) \mid M ,$$

where

a) $M \in SL(2, K)$,

b) ω *is a basis element of degree* m *(3.1).*

The subspace

$$\mathcal{E}_0(\Gamma, m) \subset \mathcal{E}(\Gamma, m)$$

consists of all closed forms.

The image of $\mathcal{E}_0(\Gamma, m)$ in $H^m(\Gamma)$ is the so-called Eisenstein cohomology. We denote it by

$$H^m_{\text{Eis}}(\Gamma) = \text{Im}(\mathcal{E}_0(\Gamma, m) \longrightarrow H^m(\Gamma)) .$$

In the following

$$\kappa_1, \ldots, \kappa_h$$

denotes a complete set of representatives of all cusp classes.

3.5 Proposition. *Assume* $n \leq m \leq 2n - 2$. *The natural restriction map*

$$H^m(\Gamma) \longrightarrow \bigoplus_{j=1}^{h} H^m(\Gamma_{\kappa_j})$$

is surjective. We have isomorphisms

$$\mathcal{E}(\Gamma, m) \overset{\sim}{\longrightarrow} H^m_{\text{Eis}}(\Gamma) \overset{\sim}{\longrightarrow} \bigoplus_{j=1}^{h} H^m(\Gamma_{\kappa_j}) .$$

Proof. The proposition is a consequence of 3.3.   □

We now come to the border case $m = 2n - 1$. In this case $H^m(\Gamma_\infty)$ is one-dimensional and generated by

$$\omega = \frac{dz_1 \wedge \ldots \wedge dz_{n-1} \wedge d\bar{z}_1 \wedge \ldots \wedge d\bar{z}_{n-1}}{y_1 \cdots y_{n-1}} \wedge dz_n .$$

We choose matrices $M_j \in SL(2, K)$ with

$$M_j \kappa_j = \infty \quad (1 \leq j \leq h) .$$

We then consider the differential forms

$$E^{(j)} := E^{M_j \Gamma M_j^{-1}}(\omega) \mid M_j .$$

We have (see 3.4)

$$\mathcal{E}(\Gamma, m) = \sum_{j=1}^{h} CE^{(j)} \quad (m = 2n - 1) .$$

The Fourier expansion of $E^{(j)}$ has the form

$$E^{(j)} = A_j + B_j/Ny + \text{higher terms} .$$

The same argument as in the case $m \leq 2n - 2$ shows

$$A_j \neq 0 \quad \Longleftrightarrow \quad \kappa_j = \infty .$$

Hence the forms $E^{(j)}$ are linearly independent and we obtain

$$\dim \mathcal{E}(\Gamma, 2n - 1) = h .$$

But in contrast to the case $m < 2n - 1$ the constant B_j is not zero!

The reader who goes carefully through the formulae of §4 will see that B_j is different from 0. But we do not need this, because we can use another type of argument: If all the B_j were 0, the proof of 3.3 would show that the mapping

$$H^{2n-1}(\Gamma) \longrightarrow \bigoplus_{j=1}^{h} H^{2n-1}(\Gamma_{\kappa_j}) \cong C^h$$

was surjective. But in §5 we will see by means of Poincaré duality that

$$\dim H^{2n-1}(\Gamma) = h - 1 .$$

To get rid of the constant B_j we consider linear combinations

$$E := \sum C_j E^{(j)} .$$

Such a linear combination is closed if and only if

$$\sum_{j=1}^{h} C_j B_j = 0 .$$

We now obtain

3.6 Proposition. *Assume $m = 2n - 1$. Let*

$$\kappa_j = M_j^{-1}(\infty) \quad (M_j \in SL(2, K) , \ 1 \leq j \leq h)$$

be a set of representatives of the cusp classes of Γ. *A linear combination*

$$E = \sum_{j=1}^{h} C_j E^{(j)} ,$$

where

$$E^{(j)} = E^{M_j \Gamma M_j^{-1}}(\omega) \mid M_j$$

$$= A_j + B_j/Ny + \text{higher terms} ,$$

is a closed form if and only if

$$\sum_{j=1}^{h} C_j B_j = 0 .$$

In this case the image of E *in* $H^m(\Gamma_{\kappa_j})$ *is* $A_j \cdot \omega \mid M_j$. *We have*

$$A_j \neq 0 \quad \Longleftrightarrow \quad \kappa_j \sim \infty \bmod \Gamma .$$

3.7 Corollary. *Assume* $m = 2n - 1$. *The dimension of the image* W *of the map*

$$H^m(\Gamma) \longrightarrow \bigoplus_{j=1}^{h} H^m(\Gamma_{\kappa_j}) = \mathbf{C}^h$$

is at least $h - 1$. *We have isomorphisms*

$$\mathcal{E}_0(\Gamma, m) \overset{\sim}{\longrightarrow} H^m_{\mathrm{Eis}}(\Gamma) \overset{\sim}{\longrightarrow} W .$$

As we already mentioned we shall prove by means of Poincaré duality that

$$\dim W = h - 1 .$$

Hence the codimension of the image will turn out to be 1.

We now define the Eisenstein cohomology in the remaining cases.

3.8 Definition.

1) $$H^0_{\mathrm{Eis}}(\Gamma) = H^0(\Gamma)$$

2) $$H^m_{\mathrm{Eis}}(\Gamma) = 0$$

if $m = 2n$ *or* $1 \leq m < n$.

We shall see later that now in all cases the Eisenstein cohomology is mapped isomorphically onto the image of

$$H^m(\Gamma) \longrightarrow \bigoplus_{j=1}^{h} H^m(\Gamma_{\kappa_j}).$$

Another justification for definition 3.8 is the

Remark. One could also try to construct Eisenstein cohomology classes in the case $0 \leq m < n$ by means of symmetrization of the basis forms

$$\omega = dy_a/y_a \quad (a \subset \{1, \ldots, n-1\}).$$

As we already mentioned at the beginning of this section the series

$$E(\omega, s) = \sum | N(cz + d) |^{-s} \omega | M$$

converges only if $\mathrm{Re}\, s > 2$. We hence cannot take the limit $s \to 0$. But nevertheless this series has an analytic continuation into the whole s-plane (as a meromorphic function) and we can look at special values of s_0 where $E(\omega, s_0)$ is closed or at values s_0 where $E(\omega, s)$ has a pole and where the residue is closed. We could do this with the methods of §4. The reader who goes through the details will find that only in the case $m = 0$ a non-vanishing cohomology class can be found along those lines, $H^0(\Gamma)$ is one-dimensional and generated by the class of a constant function (0-form). This constant function can be obtained as a residue of an Eisenstein series of the type

$$\sum | N(cz + d) |^{-s}$$

at $s = 0$.

§4 Analytic Continuation of Eisenstein Series

We fix a totally real number field K and an ideal

$$0 \neq \mathfrak{q} \subset \mathfrak{o}, \quad \mathfrak{o} = \mathfrak{o}_K.$$

We consider the main congruence subgroup

$$\Gamma_K[\mathfrak{q}] = \{M \in SL(2, \mathfrak{o}) \mid M \equiv E \bmod \mathfrak{q}\}.$$

Let $\Gamma \subset SL(2, \mathbf{R})^n$ be In §3 we introduced the series

$$E_{\alpha,\beta}(z, s) = E_{\alpha,\beta}^{\Gamma}(z, s) = \sum_{\Gamma_\infty \backslash \Gamma} N(cz + d)^{-\alpha} N(c\bar{z} + d)^{-\beta} | N(cz + d) |^{-2s},$$

where $\alpha, \beta \in \mathbf{Z}^n$ are two vectors of integers with the properties

a) $$\alpha \neq \beta \, ,$$

b) $$2r := \alpha_j + \beta_j$$

is an even number, independent of j (for our purpose $r = 1$ would be sufficient). The series converges for

$$2r + \operatorname{Re} s > 2 \, .$$

In this section we want to prove that in the case $r = 1$ the limit

$$E_{\alpha,\beta}(z) = \lim_{s \to 0} E_{\alpha,\beta}(z, s)$$

exists and we will also obtain precise information about the behaviour at the cusps.

4.1 Remark. *Let $\Gamma \subset \Gamma'$ be a subgroup which contains $SL(2, \mathbf{R})^n$ as a subgroup of finite index. Then we have*

$$E_{\alpha,\beta}^{\Gamma'}(z, s) = \sum_{M \in \Gamma \backslash \Gamma'} (E_{\alpha,\beta}^{\Gamma}(z, s)|M) \, ,$$

where the operator $\cdot \mid M$ is defined by

$$(f \mid M)(z, s) = N(cz + d)^{-\alpha} N(c\bar{z} + d)^{-\beta} \mid N(cz + d) \mid^{-2s} f(Mz, s) \, .$$

This remark, which is of course trivial, allows us to reduce the question of analytic continuation to the main congruence group. From now on we assume, unless otherwise stated, that

$$\Gamma = \Gamma_K[\mathbf{q}] \, .$$

We determine a set of representatives of $\Gamma_\infty \backslash \Gamma$. A pair (c, d) of integers (in $\mathbf{o}$) is the second row of a modular matrix ($\in SL(2, \mathbf{o})$) iff it generates the unit ideal

$$(c, d) = (1) \, .$$

A necessary condition for being the second row of a matrix in $\Gamma = \Gamma_K[\mathbf{q}]$ is of course

$$c \equiv 0 \bmod \mathbf{q}, \quad d \equiv 1 \bmod \mathbf{q} \, .$$

But this condition (together with $(c, d) = (1)$) is also sufficient, because we may replace

$$\begin{pmatrix} a & b \\ c & d \end{pmatrix} \longrightarrow \begin{pmatrix} 1 & -b \\ 0 & 1 \end{pmatrix} \begin{pmatrix} a & b \\ c & d \end{pmatrix}$$

and hence assume

$$b \equiv 0 \bmod 1 - d$$
$$\Rightarrow b \equiv 0 \bmod \mathbf{q}.$$

From the relation $ad - bc = 1$ we then obtain $a \equiv 1 \bmod \mathbf{q}$.

Two matrices

$$\begin{pmatrix} a & b \\ c & d \end{pmatrix}, \begin{pmatrix} a' & b' \\ c' & d' \end{pmatrix} \in \Gamma$$

define the same coset iff there exists a unit

$$\varepsilon \in \mathfrak{o}^*, \quad \varepsilon \equiv 1 \bmod \mathbf{q}$$

such that

$$c' = \varepsilon c, \quad d' = \varepsilon d.$$

In this case we call the pairs (c, d) and (c', d') associate mod $\mathbf{q}$. We now have the explicit form of the Eisenstein series

$$E_{\alpha,\beta}(z, s) = \sum_{\substack{(c,d)=1,(c,d)\mathbf{q} \\ c \equiv 0 \bmod \mathbf{q}, d \equiv 1 \bmod \mathbf{q}}} N(cz + d)^{-\alpha} N(c\bar{z} + d)^{-\beta} \mid N(cz + d) \mid^{-2s}.$$

The subscript $(c, d)\mathbf{q}$ indicates that the summation is taken only over a set of representatives of associate pairs.

This type of Eisenstein series was considered in 1928 already by Kloosterman.[*]

(In this connection we should mention that the method of Hecke-Kloosterman could not be generalized to other modular groups, for example the important Siegel modular group. In a deep paper, Langlands developed a method which gives analytic continuation of all Eisenstein series on semisimple Lie groups. But it would be even more complicated to extract our special case from Langlands' paper than to give the direct proof following Kloosterman.)

Before we start with the analytic continuation we have to introduce more general Eisenstein series:
Assume that (besides our level $\mathbf{q}$) a further ideal $\mathbf{a}$ is given. We do not demand that $\mathbf{a}$ be integral. For two elements

$$c_0 \in \mathbf{a}, \quad d_0 \in \mathbf{a}$$

---

[*] Kloosterman actually merely considered the case $\beta = 0$. But the general case can be easily reduced to this case by means of certain simple differential operators introduced by Maaß.

we define the Eisenstein series (of level $\mathbf{q}$) as

$$G_{\alpha,\beta}(z;s;(c_0,d_0);\mathbf{a}) = G^{\mathbf{q}}_{\alpha,\beta}(z;s;(c_0,d_0);\mathbf{a}) =$$

$$\sideset{}{'}\sum_{\substack{c \equiv c_0 \bmod \mathbf{qa} \\ d \equiv d_0 \bmod \mathbf{qa},(c,d)\mathbf{q}}} N(cz+d)^{-\alpha} N(c\bar{z}+d)^{-\beta} \mid N(cz+d) \mid^{-2s}.$$

The summation is taken over a set of representatives of pairs

$$(c,d) \in K \times K, \quad (c,d) \neq (0,0), \quad c - c_0 \in \mathbf{qa}, \quad d - d_0 \in \mathbf{qa}$$

with respect to the introduced relation: Two pairs $(c,d),(c',d')$ are called associate if there exists a unit $\varepsilon \in \mathbf{o}^*, \varepsilon \equiv 1 \bmod \mathbf{q}$, with

$$c' = \varepsilon c, \quad d' = \varepsilon d.$$

We point out that no condition of coprimeness is demanded! This series converges (for $z \in \mathbf{H}^n$) if

$$2r + 2\operatorname{Re} s > 2 \quad (2r = \alpha_j + \beta_j \in 2\mathbf{Z})$$

and represents an analytic function on s (we are interested in the case $\alpha + \beta = (2,\ldots,2)$).

4.2 Lemma. *The group*

$$Gl_+(2,K) = \{A \in GL(2,K) \mid \det A > 0\}$$

acts on the R-vector space generated by all $G_{\alpha,\beta}$ by means of the formula

$$f(z,s) \longmapsto (\det A)^{2r+2s} \cdot f(Az,s) N(cz+d)^{-\alpha} N(c\bar{z}+d)^{-\beta} \mid N(cz+d) \mid^{-2s}.$$

Proof. It is not difficult to find an explicit expression for the transformed Eisenstein series. For the sake of simplicity we make use of the simple fact: The group $Gl_+(2,K)$ is generated by the special matrices

$$\begin{pmatrix} 1 & a \\ 0 & 1 \end{pmatrix}, \begin{pmatrix} \varepsilon & 0 \\ 0 & 1 \end{pmatrix}, \varepsilon > 0, \begin{pmatrix} 0 & 1 \\ -1 & 0 \end{pmatrix}.$$

Because of the formula

$$\begin{pmatrix} \varepsilon & 0 \\ 0 & 1 \end{pmatrix} \begin{pmatrix} 1 & a \\ 0 & 1 \end{pmatrix} \begin{pmatrix} \varepsilon^{-1} & 0 \\ 0 & 1 \end{pmatrix} = \begin{pmatrix} 1 & a\varepsilon \\ 0 & 1 \end{pmatrix}$$

one may even assume that a lies in a given ideal $\mathbf{a}$. Now the proof of Lemma 4.2 is very easy. For example

$$G_{\alpha,\beta}(z+a;s;(c_0,d_0);\mathbf{a}) = G_{\alpha,\beta}(z;s;(c_0,d_0+c_0a);\mathbf{a})$$

$$G_{\alpha,\beta}(-z^{-1}; s; (c_0, d_0); \mathbf{a}) = N(z)^\alpha N(\bar{z})^\beta \mid N(z) \mid^{2s} G_{\alpha,\beta}(z; s; (d_0, -c_0); \mathbf{a}).\square$$

Our next goal is to express the Eisenstein series E as a linear combination of the G's.

For this purpose we need the notion of a "ray class mod $\mathbf{q}$".

Notation:

$\mathcal{I} = $ group of all ideals of K,

$\mathcal{H} = $ group of all principal ideals.

A (not necessarily integral) ideal $\mathbf{a} \in \mathcal{I}$ is called coprime to $\mathbf{q}$, if no prime divisor of $\mathbf{q}$ occurs in the prime decomposition of $\mathbf{a}$.

We denote by

$$\mathcal{I}(\mathbf{q}) \subset \mathcal{I}$$

the subgroup of all ideals which are coprime to $\mathbf{q}$. We also have to define a certain subgroup $\mathcal{H}(\mathbf{q})$ of the group $\mathcal{H}$ of principal ideals: A principal ideal belongs to $\mathcal{H}(\mathbf{q})$ if and only if it has a generator α with the following two properties:

a) $\alpha > 0$ (totally positive).
b) The denominator of the ideal $(\alpha - 1)\mathbf{q}^{-1}$ is coprime to $\mathbf{q}$, i.e.

$$\alpha - 1 \in \mathbf{q} \cdot \mathbf{b}, \quad \mathbf{b} \in \mathcal{I}(\mathbf{q}).$$

The usual proof of the finiteness of the class number $h = \#\mathcal{I}/\mathcal{H}$ also shows that the group

$$\mathcal{I}(\mathbf{q})/\mathcal{H}(\mathbf{q})$$

is finite.

Its elements are the so-called **ray classes** mod $\mathbf{q}$.

We also need the Möbius function $\mu(\mathbf{a})$ which is defined on the set of integral ideals. Let

$$\mathbf{a} = \mathbf{p}_1^{\nu_1} \cdot \ldots \cdot \mathbf{p}_n^{\nu_n}$$

be the prime decomposition of an integral ideal $\mathbf{a}$. One defines

$$\mu(\mathbf{a}) = \begin{cases} (-1)^n & \text{if all } \nu_i = 1 \\ 0 & \text{otherwise,} \end{cases}$$

$$\mu(\mathbf{o}) = 1.$$

The Möbius function has the basic property

$$\sum_{\mathbf{a}|\mathbf{q}} \mu(\mathbf{a}) = \begin{cases} 1 & \text{if } \mathbf{q} = \mathbf{o} \\ 0 & \text{if } \mathbf{q} \neq \mathbf{o}. \end{cases}$$

After these preparations we can give an explicit expression of E as linear combination of the G's.

Introducing the Möbius function we may get rid of the condition of coprimeness in the definition of E, namely

$$E_{\alpha,\beta}(z,s) = \sum_{\substack{c\equiv 0 \bmod \mathbf{q} \\ d\equiv 1 \bmod \mathbf{q}, (c,d)\mathbf{q}}} \sum_{\mathbf{a}|(c,d)} \mu(\mathbf{a})N(cz+d)^{-\alpha}N(c\overline{z}+d)^{-\beta} \mid N(cz+d)\mid^{-2s}.$$

The occuring ideals $\mathbf{a}$ are of course coprime with $\mathbf{q}$ (because $d\equiv 1$ mod $\mathbf{q}$). We obtain

$$E_{\alpha,\beta}(z,s) = \sum_{\substack{\mathbf{a} \text{ integral} \\ \text{coprime with } \mathbf{q}}} \mu(\mathbf{a})\cdot$$

$$\sum_{\substack{(c,d)\equiv(0,1) \bmod \mathbf{q} \\ (c,d)\equiv(0,0) \bmod \mathbf{a}, (c,d)\mathbf{q}}} N(cz+d)^{-\alpha}N(c\overline{z}+d)^{-\beta} \mid N(cz+d)\mid^{-2s}.$$

We now fix a ray class mod $\mathbf{q}$

$$\mathcal{A} \in \mathcal{I}(\mathbf{q})/\mathcal{H}(\mathbf{q})$$

and consider the contribution of this ray class to $E_{\alpha,\beta}(z,s)$:

$$E_{\alpha,\beta}(\mathcal{A};z,s) =$$

$$\sum_{\substack{\mathbf{a}\in\mathcal{A} \\ \mathbf{a} \text{ integral}}} \mu(\mathbf{a})\cdot$$

$$\sum_{\substack{(c,d)\equiv(0,1) \bmod \mathbf{q} \\ (c,d)\equiv(0,0) \bmod \mathbf{a}, (c,d)\mathbf{q}}} N(cz+d)^{-\alpha}N(c\overline{z}+d)^{-\beta} \mid N(cz+d)\mid^{-2s}.$$

Of course we have

$$E_{\alpha,\beta}(z,s) = \sum_{\mathcal{A}\in\mathcal{I}(\mathbf{q})/\mathcal{H}(\mathbf{q})} E_{\alpha,\beta}(\mathcal{A};z,s).$$

Now we fix an integral ideal in our given ray class $\mathcal{A}$:

$$\mathbf{a}_0 \in \mathcal{A}, \quad \mathbf{a}_0 \subset \mathbf{o}.$$

Then every other ideal $\mathbf{a}\in\mathcal{A}$ is of the form

$$\mathbf{a} = \gamma\cdot\mathbf{a}_0, \quad \gamma > 0.$$

$$E_{\alpha,\beta}(\mathcal{A};z,s) = \sum_{\substack{\mathbf{a}\in\mathcal{A} \\ \mathbf{a}\ \text{integral}}} \mu(\mathbf{a})N(\gamma)^{-2(r+s)}$$

$$\cdot \sum_{\substack{(c',d')\equiv(0,1)\ \text{mod}\ \mathbf{q} \\ (c',d')\equiv(0,0)\ \text{mod}\ \mathbf{a}_0,\,(c',d')\mathbf{q}}} N(c'z+d')^{-\alpha}N(c'\bar{z}+d')^{-\beta}\mid N(c'z+d')\mid^{-2s}\cdot$$

The ideals $\mathbf{q}$ and $\mathbf{a}_0$ being coprime, we can find a pair $c_0 \ (= 0)$, d_0 with the property

$$(c_0, d_0) \equiv (0,1)\ \text{mod}\ \mathbf{q}$$
$$(c_0, d_0) \equiv (0,0)\ \text{mod}\ \mathbf{a}_0\ .$$

We now obtain

$$E_{\alpha,\beta}(\mathcal{A};z,s) = \mathcal{N}(\mathbf{a}_0)^{2(r+s)} \sum_{\substack{\mathbf{a}\in\mathcal{A} \\ \mathbf{a}\ \text{integral}}} \mu(\mathbf{a})\mathcal{N}(\mathbf{a})^{-2(r+s)}\cdot G_{\alpha,\beta}(z;s;(c_0,d_0);\mathbf{a}_0).$$

4.3 Lemma. *Let* $\mathbf{m} \subset K$ *be a lattice and* $\Lambda \subset \mathbf{o}^*$ *be a subgroup of finite index acting on* $\mathbf{m}$,

$$\Lambda \times \mathbf{m} \longrightarrow \mathbf{m}$$
$$(\varepsilon, a) \longmapsto \varepsilon a\ .$$

Then if a *runs over a complete system of representatives of* $\mathbf{m} - \{0\}$ *mod* Λ *the series*

$$\sum \mid N(a) \mid^{-\sigma}, \quad \sigma > 1,$$

converges.

4.3₁ Corollary. *The series*

$$\sum_{\substack{\mathbf{a}\in\mathcal{A} \\ \mathbf{a}\ \text{integral}}} \mu(\mathbf{a})\mathcal{N}(\mathbf{a})^{-s}$$

defines an analytic function on the domain $\operatorname{Re} s > 1$.

Proof. We can choose the system of representatives such that $(a_1,\ldots,a_n)$ is contained in a fundamental domain Q of Λ acting on $\mathbf{R}^n$ by

$$(x,\varepsilon) \longmapsto x\varepsilon\ .$$

Such fundamental domains have been determined. The series can then be compared with the integral

$$\frac{1}{\text{vol}(\mathbf{m})} \int_{x\in Q,\,|N(x)|\geq 1} \mid N(x) \mid^{-\sigma} dx_1 \ldots dx_n\ . \qquad \square$$

This proof gives a little more than stated in 4.3, namely

4.3₂ Remark. *(Notations as in 4.3)*

The limit

$$\lim_{\sigma \to 1}(\sigma - 1) \sum{}' \mid N(a) \mid^{-\sigma}$$

exists (and is unequal to zero).

For our purposes we do not need the deeper result of Hecke that

$$(s - 1) \sum{}' \mid N(a) \mid^{-s}$$

has an analytic continuation as entire function into the whole s-plane.

Analytic Continuation of the Eisenstein Series G (as functions of s). We first consider the simpler series

$$f_{\alpha,\beta}(z; s; \mathbf{m}) = \sum_{g \in \mathbf{m}} N(z + g)^{-\alpha} N(\overline{z} + g)^{-\beta} \mid N(z + g) \mid^{-2s}$$

where $\mathbf{m} \subset K$ is any lattice, for example an ideal. The function $f_{\alpha,\beta}$ remains unchanged if we replace

$$z \longrightarrow z + a, \quad a \in \mathbf{m},$$

and hence admits a Fourier expansion

$$f_{\alpha,\beta}(z; s; \mathbf{m}) = \sqrt{d(\mathbf{m})} e^{(\pi i/2)S(\beta-\alpha)} \sum_{g \in \mathbf{m}^*} h_g(y) e^{2\pi i S(gx)}.$$

Here $\mathbf{m}^*$ denotes the dual lattice of $\mathbf{m}$. The square root of the discriminant $d(\mathbf{m})$ equals the volume of a fundamental parallelotope P of $\mathbf{m}$. The Fourier integral gives the following expression for h_g:

$$h_g(y) = e^{(\pi i/2)S(\alpha-\beta)} \cdot \int_P f_{\alpha,\beta}(z, s; \mathbf{m}) e^{-2\pi i S(gx)} dx$$

$$= e^{(\pi i/2)S(\alpha-\beta)} \int_{\mathbf{R}^n} N(z)^{-\alpha} N(\overline{z})^{-\beta} \mid N(z) \mid^{-2s} e^{-2\pi i S(gx)} dx$$

$$= \mid Ny \mid^{1-2s} / N(y^{\alpha+\beta}) \cdot$$
$$\int_{\mathbf{R}^n} N(1 - ix)^{-\alpha} N(1 + ix)^{-\beta} \mid N(1 - ix) \mid^{-2s} e^{-2\pi i S(gyx)} dx.$$

The integral splits into a product of n integrals of one variable. We first collect simple properties of this one-variable integral.

4.4 Lemma. *Put*

$$\alpha, \beta \in \mathbf{Z}, \quad s \in \mathbf{C},$$

such that

$$\alpha + \beta + \operatorname{Re} s > 1.$$

Then the integral

$$h(y; \alpha + s; \beta + s) := \int_{-\infty}^{\infty} (1 - it)^{-\alpha} (1 + it)^{-\beta} \mid 1 - it \mid^{-2s} e^{-ity} dt$$

converges (for arbitrary $y \in \mathbf{R}$). It has an analytic continuation as a mero-morphic function into the whole s-plane, in fact as an entire function if $y \neq 0$. Special values of h:

a) $y = 0$:

$$h(0; \alpha + s; \beta + s) = \frac{2\pi \Gamma(\alpha + \beta + 2s - 1)}{\Gamma(\alpha + s)\Gamma(\beta + s)} 2^{1-(\alpha+\beta+2s)}.$$

b) $s = 0$ ($\alpha, \beta \in \mathbf{Z}$). One has for $y > 0$

$$h(y; \alpha; \beta) = h(-y; \beta; \alpha) = e^{-y} P_{\alpha,\beta}(y),$$

where $P_{\alpha,\beta}(y)$ is a certain polynomial in y which can be computed explicitly, for example

$$P_{\alpha,\beta}(y) = 0 \quad \text{if } \alpha \le 0$$

$$P_{\alpha,0}(y) = \frac{2\pi}{(\alpha - 1)!} y^{\alpha - 1} \quad \text{if } \alpha \ge 1.$$

Basic estimate for h: If s varies in a compact set of the s-plane, there exists a constant C such that

$$\mid h(y; \alpha + s; \beta + s) \mid \le C e^{-|y|/2}.$$

Proof. If we replace t by $-t$, we observe

$$h(y; \alpha + s; \beta + s) = h(-y; \beta + s; \alpha + s)$$

and hence assume $y \ge 0$. For the computation of the integral at $y = 0$ and for the analytic continuation as well as for the basic estimate we may assume $\beta = 0$, because the integral only depends on $\alpha + s$ and $\beta + s$.

I Computation of the Integral at y = 0. Integration by parts gives

$$h(0; \alpha + s; s) = s/(\alpha - 1) \cdot [h(0; \alpha - 2 + s + 1; s + 1) - h(0; \alpha - 1 + s + 1; s)],$$

if $\alpha \neq 1$. The same recursion formula is satisfied by the Γ-expression in 4.4. It is therefore sufficient to treat the cases $\alpha = 0$ and $\alpha = -1$. In both cases the transformation

$$t^2 + 1 = x^{-1}$$

reduces the integral to an ordinary B-integral.

II Analytic Continuation.

The analytic continuation will follow from a deformation of the path of integration. We hence define the integrand

$$(1 - it)^{-\alpha}(1 + t^2)^{-s}e^{-ity}, \quad y > 0,$$

not only for real but also for complex arguments t, $\operatorname{Im} t \leq 0$. (It looks promising to deform the path of integration into the lower half-plane, because e^{-ity} is rapidly decreasing if $\operatorname{Im} t \to -\infty$). The only problem is the definition of the complex power

$$(1 + t^2)^{-s} = e^{-s\log(1+t^2)} .$$

We define

$$\log(1 + t^2) = \log |\, 1 + t^2 \,| + i \arg(1 + t^2) ,$$

where

$$\arg(1 + t^2) := \arg(t + i) + \arg(t - i) ,$$
$$-\pi/2 \leq \arg(t + i) < 3\pi/2 ,$$
$$-3\pi/2 < \arg(t - i) \leq \pi/2 .$$

This definition has the following three properties:
1) $\arg(1 + t^2) = 0$ if $t \in \mathbf{R}$.
2) $\arg(1 + t^2)$ is continuous on the domain

$$\{t \in \mathbf{C} \mid \operatorname{Im} t \leq 0 , \ it \notin [1, \infty)\} .$$

3) Let t be a point on the critical line:

$$1 < it < \infty \quad (it \in \mathbf{R}) .$$

We have

$$\lim_{u \to t, \operatorname{Re} u > 0} \arg(1 + u^2) = -\pi$$

and

$$\lim_{u \to t, \operatorname{Re} u < 0} \arg(1 + u^2) = \pi .$$

We already mentioned that our integrand is rapidly decreasing if $\operatorname{Im} t \to -\infty$. Hence we may deform the path of integration (real axis from $-\infty$ to $+\infty$).

We now decompose our integral into two parts:

a) The integral along the circle around $-i$ is of course an entire function of s.

b) We compute the jump of the integrand at the critical line if we pass it from the right to the left half-plane: The jump of the function

$$(1 + t^2)^{-s}$$

at a point t on the critical axis ($it \in]1, \infty)$) is

$$(1 + \mid t \mid^2)^{-s} [e^{\pi i s} - e^{-\pi i s}] = 2i \sin \pi i s \cdot (1 + \mid t \mid^2)^{-s} .$$

The contribution of the two vertical lines to our integral hence is

$$2i \sin \pi i s \cdot \int (1 - it)^{-\alpha} (1 + \mid t \mid^2)^{-s} e^{-ity} dt ,$$

where the path of integration is the vertical line on the right hand side (starting from a point it_0 , $t_0 > 0$). This integral again defines an entire function of s.

III The basic estimate is an immediate consequence of the formula which defines the analytic continuation.

IV The special value of h at $s = 0$: From the residue theorem we obtain for positive y

$$h(y; \alpha; \beta) = -2\pi i \operatorname*{Res}_{t=-i} (1 - it)^{-\alpha} (1 + it)^{-\beta} e^{-ity} .$$

The residue is zero if $\alpha \leq 0$. If $\alpha \geq 1$ it is $(-i)^{-\alpha}$ times the Taylor coefficient $a_{\alpha-1}$ in the expansion

$$(1 + it)^{-\beta} e^{-ity} = \sum_{\nu=0}^{\infty} a_\nu (t + i)^\nu .$$

All these Taylor coefficients are obviously products of e^{-y} with certain polynomials in y. Their trivial computation completes the proof of 4.4.   □

As a consequence of Lemma 4.4 we obtain the analytic continuation of the series $f_{\alpha,\beta}$:

4.5 Proposition. *Let* $\mathbf{m} \subset K$ *be a lattice in* K. *The series*

$$f_{\alpha,\beta}(z; s; \mathbf{m}) = \sum N(z+g)^{-\alpha} N(\bar{z}+g)^{-\beta} \mid N(z+g) \mid^{-2s}$$

has the Fourier expansion

$$\mathrm{vol}(P)e^{(\pi i/2)S(\beta-\alpha)} \cdot \sum_{g \in \mathbf{m}^*} h_g(y)e^{2\pi i S(gz)} \, ,$$

where

$$h_g(y) = \mid Ny \mid^{1-2s-2r} \prod_{j=1}^{n} h_g(2\pi g_j y_j; \alpha_j + s, \beta_j + s) \, .$$

This Fourier series defines an analytic continuation of $f_{\alpha,\beta}(z, s; \mathbf{m})$ *as meromorphic function into the whole s-plane. The only poles come from the zero Fourier coefficient, i.e.*

$$f_{\alpha,\beta}(z; s; \mathbf{m}) - \mathrm{vol}(P)e^{(\pi i/2)S(\beta-\alpha)} \cdot \mid Ny \mid^{1-2s-2r} (2\pi)^n$$

$$\cdot \prod_{j=1}^{n} \left[\frac{\Gamma(2r + 2s - 1) \cdot 2^{1-2(r+s)}}{\Gamma(\alpha_j + s)\Gamma(\beta_j + s)} \right]$$

is an entire function of s.

(Recall: $2r := \alpha_j + \beta_j \in 2\mathbf{Z}$)

We now express the Eisenstein series

$$G_{\alpha;\beta}(z; s; (c_0, d_0); \mathbf{a}) :=$$

$$\sideset{}{'}\sum_{\substack{c \equiv c_0 \bmod \mathbf{qa} \\ d \equiv d_0 \bmod \mathbf{qa}, (c,d)\mathbf{q}}} N(cz+d)^{-\alpha} N(c\bar{z}+d)^{-\beta} \mid N(cz+d) \mid^{-2s}$$

by means of the function $f_{\alpha,\beta}$.

The contribution of all pairs (c, d) with $c = 0$ is zero if $c_0 \notin \mathbf{qa}$ and

$$\sideset{}{'}\sum_{\substack{d \equiv d_0 \bmod \mathbf{qa} \\ d\mathbf{q}}} N(d)^{-2r} \mid N(d) \mid^{-2s}$$

if $c_0 \in \mathbf{qa}$. The summation is taken over a set of representatives of all

$$d \equiv d_0 \bmod \mathbf{qa}, \quad d \neq 0 \, ,$$

with respect to the "associate relation": Two elements d, d' are called associate mod $\mathbf{q}$ if there is a unit ε, $\varepsilon \equiv 1 \bmod \mathbf{q}$, with $d' = \varepsilon d$. If we introduce the number

$$\delta = \delta(c_0, \mathbf{qa}) = \begin{cases} 1 & \text{if } c_0 \in \mathbf{qa} \\ 0 & \text{elsewhere}, \end{cases}$$

we obtain

$$
G_{\alpha,\beta}(z;s;(c_0,d_0);\mathbf{a}) = \delta \cdot \sum_{\substack{d\equiv d_0 \bmod \mathbf{qa} \\ d\mathbf{q}}} |N(d)|^{-2(r+s)}
$$

$$
+ \sum_{\substack{c\equiv c_0 \bmod \mathbf{qa} \\ c\mathbf{q}}}' f_{\alpha,\beta}(cz+d_0;s;\mathbf{qa})\,.
$$

We now replace the $f_{\alpha,\beta}$ by their Fourier expression computed in 4.5. The volume of a fundamental parallelotope of $\mathbf{m}=\mathbf{qa}$ is

$$
\mathrm{vol}(P) = \sqrt{|d(\mathbf{qa})|} = \mathcal{N}(\mathbf{qa})d_K\,,
$$

where d_K denotes the discriminant of K. From 4.5 we obtain

$$
\sum_{\substack{c\equiv c_0 \bmod \mathbf{qa} \\ c\mathbf{q}}}' f_{\alpha,\beta}(cz+d_0,s;\mathbf{qa}) = \mathcal{N}(\mathbf{qa})d_K e^{(\pi i/2)S(\beta-\alpha)} Ny^{1-2s-2r}.
$$

$$
\sum_{\substack{c\equiv c_0 \bmod \mathbf{qa} \\ c\mathbf{q}}}' \sum_{g\in(\mathbf{qa})^\bullet} |Nc|^{1-2s-2r} \cdot \prod_{j=1}^{n} h(2\pi g_j c_j y_j; \alpha_j+s; \beta_j+s) e^{2\pi i S(g(cz+d_0))}\,.
$$

If we collect in $G_{\alpha,\beta}$ all terms with fixed cg we obtain the Fourier expansion

$$
G_{\alpha,\beta}(z;s;(c_0,d_0);\mathbf{a}) = \sum_{g\in\mathbf{q}^\bullet} a_g(y,s) e^{2\pi i S(gx)}\,,
$$

where the Fourier coefficients are given by the following formulae

a) $g=0$:

$$
a_0(y,s) = \delta \cdot \prod_{j=1}^{n} \frac{\Gamma(2r+2s-1)\cdot 2^{1-2(r+s)}}{\Gamma(\alpha_j+s)\Gamma(\beta_j+s)} \cdot \sum_{\substack{c\equiv c_0 \bmod \mathbf{qa} \\ c\mathbf{q}}}' |Nc|^{1-2s-2r}
$$

$$
\cdot \sum_{\substack{d\equiv d_0 \bmod \mathbf{qa} \\ d\mathbf{q}}}' |N(d)|^{-2(r+s)} + (2\pi)^n \mathcal{N}(\mathbf{qa})d_K e^{(\pi i/2)S(\beta-\alpha)} Ny^{1-2s-2r}\,.
$$

b) $g\neq 0$:

$$
a_g(y,s) = \mathcal{N}(\mathbf{qa})d_K e^{(\pi i/2)S(\beta-\alpha)} \sum_{\substack{g=c\cdot d, d\in(\mathbf{qa})^\bullet \\ c\equiv c_0 \bmod \mathbf{qa}, c\mathbf{q}}} |Nc|^{1-2s-2r}
$$

$$
\cdot e^{2\pi i S(d\cdot d_0)} \cdot Ny^{1-2s-2r} \cdot \prod_{j=1}^{n} h(2\pi g_j y_j; \alpha_j+s; \beta_j+s)\,.
$$

The sum is a finite one which can be estimated by a constant times a suitable power of $|Ny|$. So the interchange of the summation is justified. We now obtain:

4.6 Proposition. *The difference of the Eisenstein series and its zero Fourier coefficient*

$$G_{\alpha,\beta}(z; s; (c_0, d_0); \mathbf{a}) - a_0(y, s)$$

has an analytic continuation as entire function of s into the whole s-plane.

Remark. If one makes use of the fact (which we did not prove) that the series

$$\sum_{\substack{d \equiv d_0 \bmod qa \\ d \mathfrak{q}}}' |N(d)|^{-s}$$

admits an analytic continuation as meromorphic function into the whole s-plane, we obtain: The series $G_{\alpha,\beta}$, $E_{\alpha,\beta}$ admit analytic continuations into the whole s-plane as meromorphic functions.

We now assume

$$2r = \alpha_j + \beta_j = 2 \quad \text{and} \quad \alpha \neq \beta .$$

We want to investigate the Eisenstein series $G_{\alpha,\beta}$ if we approach the border of absolute convergence $s = 0$. Because we assume $\alpha \neq \beta, r = 1$, we have

$$\alpha_j \leq 0 \quad \text{or} \quad \beta_j \leq 0$$

for at least one j. This implies that

$$\frac{1}{\Gamma(\alpha_j + s)} \quad \text{or} \quad \frac{1}{\Gamma(\beta_j + s)}$$

has a zero at $s = 0$. On the other hand the limit

$$\lim_{\sigma \to 1} (\sigma - 1) \sum_{\substack{c \equiv c_0 \bmod qa \\ c \mathfrak{q}}}' |Nc|^{-\sigma}$$

exists (4.3_2).

From our explicit formula for the zero Fourier coefficient we now obtain

$$\lim_{s \to 0} a_0(y, s) = A + B/Ny ,$$

where A and B are constants.

We collect the properties of the constants A and B which were needed in §3.

4.7 Lemma. *We have*

$$\lim_{s \to 0} a_0(y, s) = A + B/Ny$$

with certain real numbers A, B. The constant B is zero if the number of all j with $\alpha_j = \beta_j = 1$ is less or equal $n - 2$.

From the Fourier expansion of $G_{\alpha,\beta}$ and especially from the results 4.4 about the function $h(y; \alpha; \beta)$ we now obtain:

4.8 Theorem. *Let α, β be two vectors of integers such that*

$$\alpha \neq \beta \quad and \quad \alpha_j + \beta_j = 2 \quad (1 \leq j \leq n).$$

The limit

$$\lim_{s \to 0} G_{\alpha,\beta}(z; s; (c_0, d_0); \mathbf{a})$$

exists and has a Fourier expansion of the following type:

$$A + B/Ny + \sum_{g \in \mathfrak{q}^*, g \neq 0} a_g P_{\alpha,\beta}(gy) e^{-2\pi S(|g|y)} e^{2\pi i S(gz)}$$

with

$$|g| := (|g_1|, \ldots, |g_n|).$$

The coefficients $a_g \in \mathbb{C}$ can be estimated by the product of a constant and a suitable power of $|Ng|$. The functions

$$y \longmapsto (Ny) \cdot P_{\alpha,\beta}(y)$$

are certain polynomials.

We do not need the explicit form of the coefficient function $P_{\alpha,\beta}(y)$. We only notice that the calculation of the special values of $h(y; \alpha; 0) = h(-y; 0; \alpha)$ in 4.4 shows

4.8₁ Remark. *Assume $\beta_j = 0$ for some j. Then $P_{\alpha,\beta}(y)$ does not depend on the variable y_j and moreover*

$$P_{\alpha,\beta}(y) = 0 \quad if \quad y_j < 0 \quad (\beta_j = 0).$$

4.8₂ Corollary. *Assume $n \geq 2$. The Eisenstein series*

$$\lim_{\sigma \to 0} \sum_{\substack{c \equiv c_0 \bmod \mathfrak{q}\mathbf{a} \\ d \equiv d_0 \bmod \mathfrak{q}\mathbf{a}, (c,d)\mathfrak{q}}} N(cz + d)^{-2} |N(cz + d)|^{-2\sigma}$$

is a holomorphic function of z.

4.9 Theorem. *Let* $\Gamma \subset SL(2,\mathbf{R})^n$ *be any congruence group with respect to the Hilbert modular group of a totally real field* K. *Let* $\alpha, \beta \in \mathbf{Z}^n$ *be two vectors of integers with the properties*

a) $\alpha_j + \beta_j = 2 \ (1 \le j \le n)$,

b) $\alpha \ne \beta$.

Then the limit

$$E_{\alpha,\beta}(z) := \lim_{s \to 0} \sum_{\Gamma_\infty \backslash \Gamma} N(cz+d)^{-\alpha} N(c\bar{z}+d)^{-\beta} |N(cz+d)|^{-2s}$$

exists. If $M \in GL(2,K)$ *is a matrix with totally positive determinant, the function*

$$(E_{\alpha,\beta}|M)(z) = N(cz+d)^{-\alpha} N(c\bar{z}+d)^{-\beta} \cdot E_{\alpha,\beta}(Mz)$$

has a Fourier expansion of the following type

$$(E_{\alpha,\beta}|M)(z) = A + B/Ny + \sum_{g \in \mathfrak{t}^0} a_g P(gy) e^{-2\pi S(|g|y)} e^{2\pi i S(gx)},$$

where A, B *denote real numbers, the function*

$$y \longmapsto (Ny) \cdot P(y)$$

is a polynomial and the numbers a_g *have moderate growth, i.e. they can be estimated by the product of a constant and a suitable power of* $|Ng|$. *The constant* B *is zero if the number of all* j *with* $\alpha_j = \beta_j = 1$ *is less or equal* $n-2$.

The number A *is* 1 *if* M *is the unit matrix but zero if the cusp* $M^{-1}(\infty)$ *is not equivalent to* ∞. *We furthermore have:*

Assume $\beta_j = 0$ *for some* j. *Then the function*

$$E_{\alpha,\beta}(z) - B/Ny$$

is holomorphic in z_j.

Proof. We only have to put together what we did in this section: We expressed $E_{\alpha,\beta}$ as a sum of $G_{\alpha,\beta}$ (with real coefficients). We proved that the group $GL(2,K)$ acts on the space which is generated by the $G_{\alpha,\beta}$ over $\mathbf{R}$ (4.2). Up to the statement about the constant A, Theorem 4.9 is hence reduced to the $G_{\alpha,\beta}$. This last statement follows from the formula

$$A = \lim_{Ny \to \infty} \lim_{\sigma \to 0} (E_{\alpha,\beta}(z,\sigma)|M)$$

$$= \lim_{\sigma \to 0} \lim_{Ny \to \infty} (E_{\alpha,\beta}(z,\sigma)|M)$$

which is easily verified for the $G_{\alpha,\beta}$ instead of $E_{\alpha,\beta}|M$ by means of the Fourier expansion. The limit

$$\lim_{Ny \to \infty} E_{\alpha,\beta}(z,\sigma)|M \quad (\sigma > 1)$$

can be computed in the same way as in the case of holomorphic Eisenstein series of weight $2r > 2$ (Chap. I, 5).

§5 Square Integrable Cohomology

The results of §3 (including §4) will allow us to write each cohomology class of $H^m(\Gamma)$ as the sum of an Eisenstein cohomology class and the class of a square integrable differential form. The latter classes can always be represented by square integrable harmonic ones. The theory of square integrable harmonic forms runs similar to the case of a compact quotient. The method developed there (§1) will give the complete determination of $H^m(\Gamma)$.

We denote by

$$H^m_{\text{squ}}(\Gamma) \subset H^m(\Gamma)$$

the subspace of all cohomology classes $[\omega']$ which can be represented by a square integrable (closed) differential form ω, i.e.

$$\omega = \omega' + d\omega''\,.$$

The form ω'' needs **not** to be square integrable.
Of course "square integrable" refers to the Poincaré metric

$$h(z) = \begin{pmatrix} y_1^{-2} & & 0 \\ & \ddots & \\ 0 & & y_n^{-2} \end{pmatrix}\,.$$

The aim of this section is the proof of the following two propositions.

5.1 Proposition. *Let*

$$\kappa_1, \ldots, \kappa_h$$

be a set of representatives of the cusp classes. The Eisenstein cohomology

$$H^m_{\text{Eis}}(\Gamma) \qquad (0 \le m \le 2n)\,,$$

defined in §3, maps isomorphically onto the image under the natural restriction map

$$H^m(\Gamma) \longrightarrow \bigoplus_{j=1}^{h} H^m(\Gamma_{\kappa_j})\,.$$

Up to now 5.1 has been proved in the cases $n \leq m \leq 2n - 2$.

5.2 Proposition. *In the case $m > 0$ we have*

$$H^m(\Gamma) = H^m_{\text{Eis}}(\Gamma) \oplus H^m_{\text{squ}}(\Gamma).$$

Remark: In the case $m = 0$ Proposition 5.2 is not true, one has

$$H^0(\Gamma) = H^0_{\text{Eis}}(\Gamma) = H^0_{\text{squ}}(\Gamma) \cong \mathbf{C}.$$

(We have $H^0_{\text{Eis}}(\Gamma) = H^0(\Gamma)$ by definition and this definition is necessary if one wants to have 5.1. On the other hand the constant form $\omega = 1$ is square integrable because $\mathbf{H}^n/\Gamma$ has finite volume with respect to the invariant measure $\omega \wedge^* \omega$. This implies $H^0(\Gamma) = H^0_{\text{squ}}(\Gamma)$).

The proof of the two propositions depends on a good knowledge of the square integrable cohomology. The latter can be investigated by means of two important general theorems about complete Riemannian manifolds (which we explain in App. III without proofs).

A) *Each square integrable cohomology class can be represented by a harmonic square integrable differential form.*
B) *Each square integrable harmonic form is closed.*
We denote by

$$\mathcal{H}^m_{\text{squ}}(\Gamma)$$

the space of all square integrable harmonic forms of degree m.

The two theorems A and B above give a surjective map

$$\mathcal{H}^m_{\text{squ}}(\Gamma) \longrightarrow H^m_{\text{squ}}(\Gamma),$$

but in contrast to the cocompact case this map need not to be injective! The space $\mathcal{H}^m_{\text{squ}}(\Gamma)$ can be determined (because of B) in precisely the same way as in the cocompact case. We only have to check which of the harmonic forms occurring in §1 are square integrable.

5.3 Lemma. *a) The universal cohomology classes (generated by $dz_i \wedge d\bar{z}_i/y_i^2$) are square integrable.*
b) Let

$$f \in [\Gamma, (2, \ldots, 2)]$$

be a (holomorphic) Hilbert modular form. The differential form

$$\omega = f(z)dz_1 \wedge \ldots \wedge dz_n$$

is square integrable if and only if f is a cusp form.

Proof. a) Up to a sign $\omega_a \wedge {}^*\omega_a$ is the invariant volume element, but H^n/Γ has finite volume.

b) One has

$$\omega \wedge {}^*\overline{\omega} = |\, f(z)\,|^2 \cdot \text{Euclidean volume element} ,$$

hence

$$\int \omega \wedge {}^*\overline{\omega} \; = \; <f,f> ,$$

where the brackets on the right denote the Petersson scalar product, introduced in Chap. II, §1. We have shown that this converges if and only if f is a cusp form.   □

5.4 Theorem. *(Compare 1.6) Let $\Gamma \subset SL(2,\mathbf{R})^n$ be a discrete subgroup such that the extended quotient of H^n/Γ is compact but such that H^n/Γ is not compact. We have a "Hodge decomposition"*

$$\mathcal{H}_{\mathrm{squ}}^m(\Gamma) = \bigoplus_{p+q=m} \mathcal{H}_{\mathrm{squ}}^{p,q}(\Gamma) ,$$

where 1) in the case $p + q \neq n$

$$\mathcal{H}_{\mathrm{squ}}^{p,q}(\Gamma) \;=\; \mathcal{H}_{\mathrm{univ}}^{p,q} \;=\; \begin{cases} \displaystyle\sum_{\#a=p} C\omega_a & \text{if } p = q \leq n \\[2mm] 0 & \text{elsewhere} \end{cases}$$

$$\left(\omega_a = \frac{dz_{a_1} \wedge d\overline{z}_{a_1}}{y_{a_1}^2} \wedge \ldots \wedge \frac{dz_{a_p} \wedge d\overline{z}_{a_p}}{y_{a_p}^2}\right),$$

2) in the case $p + q = n$

$$\mathcal{H}_{\mathrm{squ}}^{p,q}(\Gamma) \cong \mathcal{H}_{\mathrm{univ}}^{p,q} \bigoplus_{\substack{b \subset \{1,\ldots,n\} \\ \#b=q}} [\Gamma^b, (2,\ldots,2)]_0 .$$

(Recall: $[\Gamma, r]_0$ denotes the space of cusp forms.)

As a consequence of 5.4 we obtain

5.5 Lemma. *Let ω be a square integrable harmonic form of degree $m > 0$. If ∞ is a cusp of Γ, there exists a Γ_∞-invariant form α such that*

$$\omega = d\alpha .$$

5.5₁ Corollary. *Assume $m > 0$. The composite of the two mappings*

$$\mathcal{H}^m_{squ}(\Gamma) \longrightarrow H^m(\Gamma) \overset{h}{\longrightarrow} \bigoplus_{j=1} H^m(\Gamma_{\kappa_j})$$

is zero.

5.5₂ Corollary.

$$H^m_{Eis}(\Gamma) \cap H^m_{squ}(\Gamma) = \{0\} \quad \text{if } m > 0.$$

Proof. We show that a square integrable harmonic form defines the zero class in $H^m(\Gamma_\infty)$, where Γ_∞ is the stabilizer of the cusp ∞.

1) Universal classes: The forms

$$\alpha_i = \frac{dx_i}{y_i} \quad (1 \le i \le n)$$

are Γ_∞-invariant (but not Γ-invariant). One has

$$d(\alpha_i) = -\frac{dx_i \wedge dy_i}{y_i^2} = \frac{1}{2i}\omega_i \,,$$

hence

$$d(\alpha_{a_1} \wedge \omega_{a_2} \wedge \ldots \wedge \omega_{a_m}) = (1/2i)\omega_{a_1} \wedge \ldots \wedge \omega_{a_m}$$

if $m > 1$.

2) Classes coming from cusp forms: We may restrict ourselves to the case

$$f(z)dz_1 \wedge \ldots \wedge dz_n \,,$$

where

$$f(z) = \sum_{g \in t^0} a_g e^{2\pi i S(gz)}$$

is a cusp form of weight $(2, \ldots, 2)$. We have

$$a_g \ne 0 \quad \Longrightarrow \quad g > 0 \,.$$

We integrate $f(z)$ with respect to the first variable

$$g(z) = \sum a_g/(2\pi i g_1)e^{2\pi i S(gz)} \,.$$

The form

$$g(z)dz_2 \wedge \ldots \wedge dz_n$$

is Γ_∞-invariant and one has

$$d(g(z)dz_2 \wedge \ldots \wedge dz_n) = f(z)dz_1 \wedge \ldots \wedge dz_n \, . \qquad \square$$

The proof of 5.5 actually gives a little more, namely a certain approximation property:
We consider a sequence of C^∞-functions on the real line

$$\varphi_k \; : \; \mathbf{R} \longrightarrow [0,1] \, , \quad k = 1, 2, \ldots$$

with the property

$$\varphi_k(t) = \begin{cases} 1 & \text{if } t \leq k \\ 0 & \text{if } t \geq k+1 \end{cases}$$

and

$$| \, \varphi'_k(t) \, | \leq 2 \, .$$

We define

$$\phi_k \; : \; \mathbf{H}^n \; \longrightarrow \; [0,1]$$

by

$$\phi_k(z) \; = \; \varphi_k(Ny) \, .$$

In the notation of Lemma 5.5 we now consider

$$\alpha_k \; := \; \phi_k \cdot \alpha$$

and

$$\omega_k \; := \; d(\alpha_k) \, .$$

We certainly have

$$\alpha_k \; \longrightarrow \; \alpha$$
$$\omega_k \; \longrightarrow \; \omega$$

(pointwise convergence). But the explicit construction during the proof of Lemma 5.5 gives a little more, namely

5.5₃ Remark. *With the notations of Lemma 5.5 we have the following approximation result:*

$$\int_{U_C/\Gamma_\infty} \beta \wedge \omega = \lim_{k \to \infty} \int_{U_C/\Gamma_\infty} \beta \wedge \omega_k$$

$$(U_C = \{z \in \mathbf{H}^n \; | \; Ny > C\} \, , \, C > 0) \, ,$$

where β is any square integrable harmonic form of complementary degree.

We leave the proof to the reader.

For the proof of Propositions 5.1 and 5.2 we need a further tool, namely the **Poincaré duality**.

Recall (see App. III): The de Rham complex has a certain subcomplex which consists of all differential forms with compact support. The cohomology groups of this subcomplex are the cohomology groups with compact support which we denote by

$$H_c^m(\Gamma).$$

One has a natural linear mapping

$$H_c^m(\Gamma) \longrightarrow H^m(\Gamma),$$

which is in general neither injective nor surjective. Obviously

$$H_c^0(\Gamma) = 0.$$

The following two theorems are explained (but not proved) in App. III.

1) (Poincaré duality): The mapping

$$(\omega, \omega') \longmapsto \int_{\mathbf{H}^n/\Gamma} \omega \wedge \omega',$$

where ω, ω' are closed differential forms, the first one with compact support, induces a non-degenerate pairing

$$H_c^m(\Gamma) \times H^{2n-m}(\Gamma) \longrightarrow \mathbf{C}.$$

We especially have

$$\dim H_c^m(\Gamma) = \dim H^{2n-m}(\Gamma).$$

2) There exists a linear mapping

$$\delta : \bigoplus_{j=1}^{h} H^m(\Gamma_{\kappa_j}) \longrightarrow H_c^{m+1}(\Gamma)$$

such that the long sequence

$$\cdots \longrightarrow H_c^m(\Gamma) \longrightarrow H^m(\Gamma) \longrightarrow \bigoplus_{j+1}^{h} H^m(\Gamma_{\kappa_j}) \longrightarrow H_c^{m+1}(\Gamma) \longrightarrow \cdots$$

is exact.

We use this sequence in the case $m = 2n - 1$ and obtain the exact sequence

$$H^{2n-1}(\Gamma) \xrightarrow{\ h\ } \bigoplus_{j=1}^{h} H^{2n-1}(\Gamma_{\kappa_j}) \longrightarrow H_c^{2n}(\Gamma) \longrightarrow H^{2n}(\Gamma).$$

From

$$H_c^{2n}(\Gamma) \cong H^0(\Gamma) \cong \mathbb{C}$$
$$H^{2n}(\Gamma) \cong H_c^0(\Gamma) = 0$$

we obtain:

The image of

$$H^{2n-1}(\Gamma) \longrightarrow \bigoplus_{j=1}^{h} H^{2n-1}(\Gamma_{\kappa_j})$$

is a subspace of codimension 1. This completes the proof of Proposition 5.1 in the case $m = 2n - 1$.

The cases $m \leq n$ now can be treated by duality:

From the surjectivity of the restriction map

$$H^m(\Gamma) \longrightarrow \bigoplus_j H^m(\Gamma_{\kappa_j})$$

in the case $n \leq m \leq 2n - 2$ and from the long exact sequence we obtain that

$$H_c^{m+1}(\Gamma) \hookrightarrow H^{m+1}(\Gamma)$$

is injective in those cases. Dualizing this result we obtain:

5.5₄ Remark. *The map*

$$H_c^m(\Gamma) \longrightarrow H^m(\Gamma)$$

is surjective if $1 \leq m < n$.

The image of this mapping is of course contained in the square integrable cohomology. We obtain

$$H^m(\Gamma) = H_{\mathrm{squ}}^m(\Gamma) \quad \text{if } 1 \leq m < n.$$

From Lemma 5.5 we finally obtain that

$$H^m(\Gamma) \longrightarrow \bigoplus_{j=1}^{h} H^m(\Gamma_{\kappa_j})$$

is the zero mapping if $0 < m \leq n$.

This jusitifies the definition $H^m_{\text{Eis}}(\Gamma) = 0$ in these cases!

The proof of Proposition 5.1 is now complete. From 5.5_1 and from the long exact sequence we conclude furthermore that the square integrable cohomology is contained in the image of the cohomology with compact support if $m > 1$. Hence both are equal and the square integral cohomology is precisely the kernel of the restriction map (5.1) (if $m > 1$). Now Propositon 5.2 follows from 5.1.

Our next goal is to determine the kernel of the mapping

$$\mathcal{H}^m_{\text{squ}}(\Gamma) \longrightarrow H^m(\Gamma).$$

5.6 Lemma. *The natural mapping*

$$\mathcal{H}^m_{\text{squ}}(\Gamma) \longrightarrow H^m(\Gamma)$$

a) is injective if $m < 2n$

b) is the zero mapping if $m = 2n$.

This means

$$H^m_{\text{squ}}(\Gamma) \cong \begin{cases} \mathcal{H}^m_{\text{squ}}(\Gamma) & \text{if } m < 2n, \\ 0 & \text{if } m = 2n. \end{cases}$$

Proof. Let

$$\omega \in \mathcal{H}^m_{\text{squ}}(\Gamma)$$

be a square integrable harmonic form whose cohomology class in $H^m(\Gamma)$ is zero. From the existence of the Poincaré pairing it follows

$$\int \omega \wedge \alpha = 0,$$

where α is a compactly supported closed differential form of degree $2n - m$. We want to show that in the case $m < 2n$ this implies $\omega = 0$, or equivalently

$$\int \omega \wedge {}^*\bar{\omega} = 0.$$

The convergence of this integral follows from the explicit description of the square integrable harmonic forms. The idea now is to approximate ${}^*\bar{\omega}$ by compactly supported closed forms. We now apply Lemma 5.5 to ${}^*\bar{\omega}$ instead of ω. We may apply this lemma to write ${}^*\bar{\omega}$ as the derivative of a certain form in a small neighbourhood of an arbitrary cusp class. These differential forms can be glued together to one form α by means of "partition of unity". The result of this construction is a form β whith compact support such that

$\bar{\omega} - \beta = d\alpha$. By means of the approximation lemma 5.5_3 we may refine this construction:

There exists a sequence of compactly supported differential forms β_k such that

a) $^*\bar{\omega} - \beta_k = d\alpha_k$

b) $\int_{\mathbf{H}^n/\Gamma} \omega \wedge \beta_k \longrightarrow \int_{\mathbf{H}^n/\Gamma} \omega \wedge^* \bar{\omega}$.

The integrals in the sequence vanish by assumption. We obtain $\omega = 0$ as desired. $\square$

Final Remark: We now have the complete picture of the cohomology and also of cohomology with compact support (by means of Poincaré duality) and the square integrable cohomology. There is also the notation of the **cuspidal cohomology**.

Let $f : \mathbf{H}^n \longrightarrow \mathbf{C}$ be a continuous function which is periodic with respect to some lattice $\mathbf{t} \subset \mathbf{R}^n$. We call f a cusp form at ∞, if the zero Fourier coefficient

$$\int_{\mathbf{R}^n/\mathbf{t}} f(z) dx_1 \ldots dx_n$$

vanishes (this coefficient is a function of y).

A Γ-invariant differential form ω on $\mathbf{H}^n$ is called a cusp form, if all the components of

$$\omega \mid M , \quad M \in SL(2,K) ,$$

are cusp forms at ∞.

The cuspidal part

$$H^m_{\mathrm{cusp}}(\Gamma)$$

consists of all cohomology classes which may be represented by a cusp form. It can be shown that each cusp form is square integrable, hence we have

$$H^m_{\mathrm{cusp}}(\Gamma) \subset H^m_{\mathrm{squ}}(\Gamma) .$$

The universal forms ω_a, $a \subset \{1,\ldots,n-1\}$, are obviously not cusp forms. From the explicit description 5.4 we obtain

$$H^m_{\mathrm{cusp}}(\Gamma) \cong \bigoplus_{b \subset \{1,\ldots,n\}} [\Gamma^b, (2,\ldots,2)]_0$$

if $m = n$ and 0 elsewhere.

§6 The Cohomology of Hilbert's Modular Groups

We only have to collect the results of the previous sections to get a complete picture of the cohomology of the Hilbert modular group, more generally of congruence groups.

The formula in the Betti and Hodge numbers involve several invariants of those groups like volume of a fundamental domain, number of elliptic fixed points of given type and certain L-series coming from the cusps. All these invariants can be computed in case of real quadratic fields.

In the following, Γ denotes a congruence group, and $\kappa_1, \ldots, \kappa_h$ representatives of the cusp classes. In the Sects. 3,4,5 we investigated the restriction map. The most difficult part of the theory was the construction of an injective homomorphism

$$\bigoplus_{j=1}^{h} H^m(\Gamma_{\kappa_j}) \longrightarrow \text{space of } \Gamma\text{-invariant differential forms of degree } m$$

in the cases $n \leq m < 2n$. The image of this map is the **space of Eisenstein series**

$$\mathcal{E}(\Gamma, m) .$$

As the Eisenstein series did not converge absolutely, we had to do the tedious job of analytic continuation (§4). Not all the Eisenstein series are closed differential forms. The subspace of closed forms has been denoted by

$$\mathcal{E}_0(\Gamma, m) \subset \mathcal{E}(\Gamma, m) .$$

In case $m = 2n - 1$ this subspace has codimension 1.In the cases $n \leq m \leq 2n - 2$ both spaces agree.The natural map of $\mathcal{E}_0(\Gamma, m)$ into the cohomology group of Γ is injective and hence defines an isomorphism to a certain subspace

$$H^m_{\text{Eis}}(\Gamma) \subset H^m(\Gamma) .$$

The case $m < n$ could be treated by means of **Poincaré duality**.We have been forced to define

$$H^m_{\text{Eis}}(\Gamma) = 0, \quad \text{if} \quad 0 < m < n ,$$

and

$$H^0_{\text{Eis}} = H^0(\Gamma) \, (= \mathbf{C}) .$$

The main result of this construction is

6.1 Theorem. *The Eisenstein cohomology $H^m_{\text{Eis}}(\Gamma)$ maps isomorphically to the image of the mapping*

$$H^m(\Gamma) \longrightarrow \bigoplus_{j=1}^{h} H^m(\Gamma_{\kappa_j}) .$$

Its dimension is

$$
b_{\text{Eis}}^m = \begin{cases} 0 & \text{if } 0 < m < n \\ h \cdot \binom{n-1}{m-n} & \text{if } n \leq m < 2n-1 \\ h-1 & \text{if } m = 2n-1 . \end{cases} \qquad \backslash
$$

We now have to consider the space

$$
\mathcal{H}_{\text{squ}}^m(\Gamma)
$$

of all square intgrable harmonic differential forms of degree m. By a general theorem they are all closed. Hence one obtains a mapping

$$
\mathcal{H}_{\text{squ}}^m(\Gamma) \longrightarrow H^m(\Gamma) .
$$

In §5 we have proved

6.2 Theorem. *Assume $m < 2n$. The natural map*

$$
\mathcal{H}_{\text{squ}}^m(\Gamma) \longrightarrow H^m(\Gamma)
$$

is injective. Its image is the so-called square integrable part of the cohomology. We denote it by

$$
H_{\text{squ}}^m(\Gamma) .
$$

In the cases $m > 0$ it coincides with the image of the cohomology with compact support and we have in this case

$$
H^m(\Gamma) = H_{\text{Eis}}^m(\Gamma) \oplus H_{\text{squ}}^m(\Gamma) .
$$

The square integrable part of the cohomology can be determined by the method of Matsushima and Shimura (§1). One obtains a further decomposition

$$
H_{\text{squ}}^m(\Gamma) = H_{\text{univ}}^m(\Gamma) \oplus H_{\text{cusp}}^m(\Gamma) .
$$

The universal part of the cohomology is generated by the harmonic and square integrable differential forms

$$
\omega_a , \quad a \subset \{1, \ldots, n-1\} .
$$

The cuspidal part only arises in the case $m = n$. It comes from holomorphic cusp forms of weight $(2, \ldots, 2)$, which belong to certain conjugate groups of Γ:

$$
H_{\text{cusp}}^n(\Gamma) \cong \bigoplus_{b \subset \{1, \ldots, n\}} [\Gamma^b, (2, \ldots, 2)]_0 .
$$

We collect these results to obtain formulae for the Betti numbers

$$b^m = \dim_{\mathbf{Z}} H^m(\Gamma)$$

of an arbitrary congruence group.

6.3 Theorem. *The Betti numbers of an arbitrary congruence group Γ of some Hilbert modular group are given by the following formulae:*
1) $b^0 = 1$, $b^{2n} = 0$.
2) Assume $0 < m < 2n$. Then we have

$$b^m = b^m_{\text{univ}} + b^m_{\text{Eis}} + b^m_{\text{cusp}} ,$$

where

$$\text{a)} \quad b^m_{\text{univ}} = \begin{cases} \binom{n}{m/2} & \text{if } m \text{ is even} \\ 0 & \text{if } m \text{ is odd,} \end{cases}$$

$$\text{b)} \quad b^m_{\text{Eis}} = \begin{cases} 0 & \text{if } 0 < m < n \\ h \cdot \binom{n-1}{m-n} & \text{if } n \leq m < 2n - 1 \\ h - 1 & \text{if } m = 2n - 1, \end{cases}$$

$$\text{c)} \quad b^m_{\text{cusp}} = \begin{cases} 0 & \text{if } m \neq n \\ \sum_{p+q=m} h^{p,q}_{\text{cusp}} & \text{if } m = n, \end{cases}$$

where

$$h^{p,q}_{\text{cusp}} = \sum_{\substack{b \subset \{1,\ldots,n\} \\ \#b=p}} \dim[\Gamma^b, (2,\ldots,2)]_0 .$$

§7 The Hodge Numbers of Hilbert Modular Varieties

by Claus Ziegler

Introduction

Let K be a totally real algebraic numberfield of degree $n := \dim_{\mathbf{Q}} K$ and let $o \subset K$ denote the ring of algebraic integers. The Hilbert modular group $\Gamma_K := SL(2, o)$ can be regarded as a discrete subgroup of $SL(2, \mathbf{R})^n$ acting discontinuously on the n-fold product of upper half-planes $\mathbf{H}$. More generally we shall consider congruence subgroups $\Gamma \subset \Gamma_K$ defined by

$$\Gamma = \Gamma_K[\mathbf{a}] := \operatorname{Kernel}\left(SL(2, \mathbf{o}) \to SL(2, \mathbf{o}/\mathbf{a})\right)$$

where $\mathbf{a} \subset \mathbf{o}$ denotes some ideal. Γ is of finite index in Γ_K, the cusps of Γ are the elements of $K \cup \{\infty\}$. Let $(\mathbf{H}^n)^* := \mathbf{H}^n \cup K \cup \{\infty\}$ and $X_\Gamma := (\mathbf{H}^n)^*/\Gamma$ be equipped with the topologies described in Chap. I, §2, then X_Γ is a compact normal complex space. In the case $n \geq 2$ X_Γ contains a finite number of singular points, namely the classes of cusps

$$S := \{[k] \mid k \in K \cup \{\infty\}\}$$

together with the classes of elliptic fixed points

$$F := \{[z] \mid z \in \mathbf{H}^n \text{ elliptic fixed point}\}.$$

$X := X_\Gamma \backslash (S \cup F)$ is a quasi-projective complex manifold of real dimension $2n$.

Concerning the singularities of X_Γ, a resolution with the following properties can be constructed (see Ehlers [11]):

1) There exists a compact complex manifold $\overline{X}$ and a morphism of complex spaces $f := \overline{X} \to X_\Gamma$
2) $f|f^{-1}(X) : f^{-1}(X) \overset{\sim}{\longrightarrow} X$ is a biholomorphic mapping
3) f is proper
4) $Y := f^{-1}(X_\Gamma \backslash X)$ is a divisor with normal crossings in $\overline{X}$, i.e. in local co-ordinates we have $Y = \{z \mid z_1 \cdots z_k = 0, \quad 0 \leq k \leq n\}$

Especially we can consider X as an open dense subspace of $\overline{X}$.

The following results are due to Deligne (see Deligne [9]): Let X be a quasi-projective complex manifold and $\overline{X} \supset X$ a compactification of X, such that $Y := \overline{X} \backslash X$ is a divisor with normal crossings in $\overline{X}$. Let $j : X \hookrightarrow \overline{X}$ denote the natural imbedding of X into $\overline{X}$. We consider the logarithmic de Rham-complex

$$\Omega_{\overline{X}}^{\cdot}\langle Y \rangle : \qquad \ldots \to 0 \to \Omega_{\overline{X}}^0\langle Y \rangle \overset{d}{\longrightarrow} \Omega_{\overline{X}}^1\langle Y \rangle \overset{d}{\longrightarrow} \Omega_{\overline{X}}^2\langle Y \rangle \overset{d}{\longrightarrow} \ldots$$

where $\Omega_{\overline{X}}^1\langle Y \rangle \subset j_* \Omega_X^1$ denotes the locally free $\mathcal{O}_{\overline{X}}$-submodule of $j_* \Omega_X^1$ generated by $\Omega_{\overline{X}}^1$ and the differential forms $\frac{dz_i}{z_i}$ for z_i local defining equation of some irreducible local component of Y and $\Omega_{\overline{X}}^m\langle Y \rangle := \bigwedge^m \Omega_{\overline{X}}^1\langle Y \rangle$.

The complex $\Omega_{\overline{X}}^{\cdot}\langle Y \rangle$ is equipped with two natural filtrations, namely the Hodge filtration F and the weight filtration W. These are defined as follows:

$$F^p(\Omega_{\overline{X}}^m\langle Y \rangle) := \begin{cases} 0 & m < p \\ \Omega_{\overline{X}}^m\langle Y \rangle & m \geq p \end{cases}$$

and $W_l(\Omega^m_{\overline{X}}\langle Y \rangle)$ is defined to be the locally free $\mathcal{O}_{\overline{X}}$-submodule of $\Omega^m_{\overline{X}}\langle Y \rangle$ generated by differential forms of the form

$$ \alpha \wedge \frac{dz_{i1}}{z_{i1}} \wedge \ldots \wedge \frac{dz_{ik}}{z_{ik}}, \qquad k \leq l $$

with α holomorphic and the z_{ij}'s being local defining equations of distinct irreducible local components Y_i of Y.

The complex $(\Omega^{\cdot}_{\overline{X}}\langle Y \rangle, F, W)$ is biregular bifiltered and admits a canonical bifiltered acyclic resolution $(K^{\cdot}, F, W)$ (compare Deligne [9]):

$$ K^m := \sum_{p'+q'=m} \Omega^{p'}_{\overline{X}}\langle Y \rangle \otimes M^{0,q'}_{\overline{X}} $$

$$ F^p(K^m) := \sum_{p'+q'=m} F^p(\Omega^{p'}_{\overline{X}}\langle Y \rangle) \otimes M^{0,q'}_{\overline{X}} = \sum_{\substack{p'+q'=m \\ p' \geq p}} \Omega^{p'}_{\overline{X}}\langle Y \rangle \otimes M^{0,q'}_{\overline{X}} $$

$$ W_l(K^m) := \sum_{p'+q'=m} W_l(\Omega^{p'}_{\overline{X}}\langle Y \rangle) \otimes M^{0,q'}_{\overline{X}}. $$

In the following we will refer to $K^{\cdot}$ as the differentiable logarithmic de Rham-complex.

Now we obtain induced filtrations F, W on the hypercohomology groups $\mathbf{H}^m(\overline{X}, \Omega^{\cdot}_{\overline{X}}\langle Y \rangle) \simeq H^m(\Gamma K^{\cdot})$ and thus we are led to filtrations on the cohomology of X by the isomorphism $\mathbf{H}^m(\overline{X}, \Omega^{\cdot}_{\overline{X}}\langle Y \rangle) \simeq H^m(X, \mathbf{C})$. The key results of Deligne are:

1) The filtrations F, W on $H^m(X, \mathbf{C})$ are independent of the choice of the compactification $\overline{X}$.
2) The filtrations $F, W[m]$ define a mixed Hodge structure on $H^m(X, \mathbf{C})$, which depends functorially on X.

Especially we get invariants of X, the so called Hodge numbers:

$$ h^{p,q}_m(X) := \dim_{\mathbf{Z}} Gr^p_F Gr^q_{\overline{F}} Gr^{W[m]}_{p+q}(H^m(X, \mathbf{C})). $$

The classical Hodge theory is contained in Deligne's theory:

If $X = \overline{X}$ is compact, then the mixed Hodge structure on $H^m(X, \mathbf{C})$ is equivalent to the classical Hodge decomposition $H^m(X, \mathbf{C}) = \bigoplus_{p+q=m} H^{p,q}(X)$, the $h^{p,q}_m(X)$ are zero if $p+q \neq m$ and coincide with the classical Hodge numbers $h^{p,q} := \dim_{\mathbf{Z}} H^{p,q}(X)$ in the remaining case.

We want to make use of Deligne's theory to investigate the manifold $X = X_\Gamma \backslash (S \cup F)$; we restrict ourselves to the special case of Γ acting freely on $\mathbf{H}^n$. Then we have $F = \emptyset$ and $X = \mathbf{H}^n/\Gamma$. The cohomology $H^m(\mathbf{H}^n/\Gamma, \mathbf{C})$ is well known (see §5,6); if $m > 0$ we have the decomposition

$$H^m(\mathbf{H}^n/\Gamma, \mathbf{C}) = H^m(\Gamma) = H^m_{\text{univ}}(\Gamma) \oplus H^m_{\text{cusp}}(\Gamma) \oplus H^m_{\text{Eis}}(\Gamma),$$

which has been described in §6.

In this section we shall study the filtrations on $H^m(\mathbf{H}^n/\Gamma, \mathbf{C}) = H^m(\Gamma)$ existing in case of no elliptic fixed points. We first have a look at the square integrable cohomology $H^m_{\text{squ}}(\Gamma) := H^m_{\text{univ}}(\Gamma) \oplus H^m_{\text{cusp}}(\Gamma)$, before we turn our attention to the Eisenstein cohomology. Then we summarize our results and describe the Hodge numbers $h^{p,q}_m(\mathbf{H}^n/\Gamma)$. The appendix gives a short summary of the construction in Ehlers [11], emphasizing the results essential for our purposes.

Weight- and Hodge Filtration on $H^m(\Gamma)$. The logarithmic de-Rahm-complex $\Omega^{\cdot}_{\overline{X}}\langle Y \rangle$ admits different acyclic resolutions: Besides $K^{\cdot}$ (compare introduction) also $j_*M^{\cdot}_X$ is an acyclic resolution of $\Omega^{\cdot}_{\overline{X}}\langle Y \rangle$. Therefore the natural imbeddings $\Omega^m_{\overline{X}}\langle Y \rangle \hookrightarrow K^m \hookrightarrow j_*M^m_X$ induce quasi-isomorphisms

$$\Omega^{\cdot}_{\overline{X}}\langle Y \rangle \to K^{\cdot} \to j_*M^{\cdot}_X$$

Hence:

$$\mathbf{H}^m(\overline{X}, \Omega^{\cdot}_{\overline{X}}\langle Y \rangle) \simeq H^m(\Gamma K^{\cdot}) \simeq H^m(\Gamma j_*M^{\cdot}_X) \simeq$$
$$H^m(\Gamma M^{\cdot}_X) \simeq H^m(\Gamma) \simeq H^m(X, \mathbf{C})$$

Because of that the isomorphism $H^m(\Gamma K^{\cdot}) \overset{\sim}{\to} H^m(\Gamma)$ is induced by the natural cochain map $K^{\cdot} \hookrightarrow j_*M^{\cdot}_X$. We obtain:

7.1 Proposition. *Each cohomology class $[w] \in H^m(\Gamma)$ contains a logarithmically singular representative $\omega' \in K^m$.*

Remark. If $([\omega_1], \ldots, [\omega_s])$ is a basis of $H^m(\Gamma)$ represented by logarithmically singular differential forms $\omega_1, \ldots, \omega_s$, then $([\omega_1]_{\Gamma K^{\cdot}}, \ldots, [\omega_s]_{\Gamma K^{\cdot}})$ is a basis of $H^m(\Gamma K^{\cdot})$.

Weight- and Hodge filtration on $H^m(\Gamma)$ are defined as filtrations induced by $(K^{\cdot}, F, W)$ on $H^m(\Gamma K^{\cdot}) \overset{\sim}{\to} H^m(\Gamma)$, i.e.

$$W^l(H^m(\Gamma K^{\cdot})) := \frac{W^l(K^m) \cap Z^m}{W^l(K^m) \cap B^m} \; ; \qquad F^p(H^m(\Gamma K^{\cdot})) := \frac{F^p(K^m) \cap Z^m}{F^p(K^m) \cap B^m},$$

where $Z^m := \text{Kernel}(K^m \overset{d}{\longrightarrow} K^{m+1})$ and $B^m := \text{Im}(K^{m-1} \overset{d}{\longrightarrow} K^m)$. Thus a cohomology class $[\omega] \in H^m(\Gamma K^{\cdot})$ is contained in $W^l(H^m(\Gamma K^{\cdot}))$ if and only if it contains a representative $\omega' \in W^l(K^m)$. The same is true concerning the Hodge filtration.

Considering the different (p, q)-types of differential forms we obtain filtrations F and $\tilde{F}$ on $K^{\cdot}$, resp. $j_*M^{\cdot}_X$, which induce filtrations on the hypercohomology groups of these complexes. The filtration F on $H^m(\Gamma K^{\cdot}) \overset{\sim}{\to} H^m(\Gamma)$

is nothing else but the Hodge filtration, but unfortunately the natural imbedding

$$(K^{\cdot}, F) \hookrightarrow (j_* M_X^{\cdot}, \tilde{F})$$

is not a filtered quasi-isomorphism. Therefore the filtrations F and $\tilde{F}$ generally do not coincide on $H^m(\Gamma)$, i.e. the Hodge filtration on $H^m(\Gamma)$ is generally not induced by the (p, q)-grading of differential forms on $M_X^{\cdot}$.

We want to determine the weight filtration on $H^m(X, \mathbb{C}) \simeq H^m(\Gamma)$, therefore we take a look at the decomposition

$$H^m(\Gamma K^{\cdot}) \stackrel{\sim}{\rightarrow} H^m(\Gamma) \simeq H^m_{\text{squ}}(\Gamma) \oplus H^m_{\text{Eis}}(\Gamma) \quad.$$

We first consider the square integrable part of the cohomology $H^m_{\text{squ}}(\Gamma)$. The image of the natural mapping

$$H^m_{\text{squ}}(\Gamma) \longrightarrow \bigoplus_{j=1}^{h} H^m(\Gamma_{k_j})$$

is identically zero for $m > 0$, i.e. the representatives of $H^m_{\text{squ}}(\Gamma)$ are locally exact at the cusps. Let $[\omega] \in H^m_{\text{squ}}(\Gamma)$, then there exist open neighbourhoods $V_i \subset\subset U_i$ of the cusp classes $(k_i)_{i=1...h}$ with $U_i \cap U_j = \emptyset$ for $i \neq j$ and differential forms η_i on U_i, such that $d\eta_i = \omega|U_i$. Let $(\phi_i)_{i=1...k}$ be a family of C^∞-functions on X_Γ with the following properties:

1) supp $\phi_i \subset U_i$
2) $\phi_i|V_i \equiv 1$
3) $0 \leq \phi_i \leq 1$

Then the differential form $\eta = \sum_{i=1}^{h} \phi_i \eta_i$ is defined on $\mathbb{H}^n/\Gamma$, thus $\omega' := \omega - d\eta$ is a representative of $[\omega] \in H^m_{\text{squ}}(\Gamma)$. We have

$$\omega'|V_i = \omega|V_i - d\eta|V_i = \omega|V_i - d\eta_i|V_i \equiv 0 \qquad \text{for } i = 1 \ldots h,$$

i.e. we have constructed a representative ω' of $[\omega]$, which vanishes identically on an open neighbourhood $V = V_1 \cup \ldots \cup V_h$ of the cusp classes, consequently $\omega' \equiv 0$ on an open neighbourhood of the divisor Y in $\overline{X}$, i.e. $\omega' \in W^0(K^m)$, therefore $[\omega] = [\omega'] \in W^0(H^m(\Gamma))$. We obtain:

7.2 Proposition.

$$H^m_{\text{squ}}(\Gamma) \subset W^0(H^m(\Gamma))$$

To get some information about the Hodge filtration F of $H^m_{\text{squ}}(\Gamma)$ we take a look at the (p, q)-types of our constructed representatives $\omega' \in W^0(K^m)$. First we consider the canonical basis elements of the universal cohomology

$H_{\mathrm{univ}}^{m}(\Gamma)$. These are given by $\omega_I = \omega_{i1} \wedge \ldots \wedge \omega_{il}$, where

$$\omega_{ik} := \frac{dz_{ik} \wedge d\bar{z}_{ik}}{y_{ik}^2} \qquad \text{and} \qquad I = \{i_1, \ldots, i_l\} \subset \{1, \ldots, n\} \quad .$$

The differential forms $\alpha_{ik} := i \dfrac{dz_{ik}}{y_{ik}}$ are Γ_∞-invariant and satisfy:

$$d\alpha_{ik} = i\frac{\partial}{\partial \bar{z}_{ik}}\left(\frac{1}{y_{ik}}\right) dz_{ik} \wedge d\bar{z}_{ik}$$

$$= -\frac{\partial}{\partial y_{ik}}\left(\frac{1}{y_{ik}}\right) dz_{ik} \wedge d\bar{z}_{ik} = \frac{dz_{ik} \wedge d\bar{z}_{ik}}{y_{ik}^2} = \omega_{ik}$$

Consequently we have $d(\alpha_{i1} \wedge \omega_{i2} \wedge \ldots \wedge \omega_{il}) = d\alpha_{i1} \wedge \omega_{i2} \wedge \ldots \wedge \omega_{il} = \omega_{i1} \wedge \ldots \wedge \omega_{il} = \omega_I$, where $\alpha_I := \alpha_{i1} \wedge \omega_{i2} \wedge \ldots \wedge \omega_{il}$ is a differential form of type $(l, l-1)$. Transformation of an arbitrary cusp k to $\{\infty\}$ by some $G \in SL(2, K)$ shows: The differential forms η_i satisfying $d\eta_i = \omega_I$ in sufficiently small neighbourhoods U_i of the cusp classes k_i may be chosen of type $(l, l-1)$; for this one has to use the fact that $G^* \omega_I = \omega_I$ for $G \in SL(2, K) \subset SL(2, \mathbf{R})^n$. We obtain:

$$\omega_I' = \omega_I - d\left(\sum_{i=1}^{h} \phi_i \eta_i\right) \in F^l(K^{2l})$$

since $\phi_i \eta_i$ is of type $(l, l-1)$. We have shown:

7.3 Proposition.

$$H_{\mathrm{univ}}^{2l}(\Gamma) \subset F^l(H^{2l}(\Gamma))$$

Next we shall consider the cuspidal part of the cohomology $H_{\mathrm{cusp}}^n(\Gamma)$. (Since $H_{\mathrm{cusp}}^m(\Gamma) = \{0\}$ for $m \neq n$, only the case $m = n$ is interesting). The canonical basis elements are given by $\omega = f\, dz_a \wedge d\bar{z}_b$, where $a \cap b = \emptyset$, $a \cup b = \{1, \ldots, n\}$ and $f(\sigma_b(z)) \in [\Gamma^b, (2, \ldots, 2)]_0$. (see §5). First let $b \neq \emptyset$. In this case we can use an argument similar to the one given above. Since $f(\sigma_b(z)) \in [\Gamma^b, (2, \ldots, 2)]_0$, f has a fourier series expansion

$$f(z) = \sum_{g \in t^0} a_g e^{2\pi i \langle g, \sigma_b(z) \rangle}$$

where $a_g \neq 0$ only if $g > 0$. Since $b \neq \emptyset$ we can choose $\nu \in b$ and integrate $f(z)$ with respect to z_ν. We obtain:

$$g(z) = -\sum_{g \in t^0} \frac{a_g}{2\pi i g_\nu} e^{2\pi i \langle g, \sigma_b(z) \rangle}$$

(Observe that $g_\nu > 0$ for $a_g \neq 0$, so this expression makes sense).

The differential form $\alpha := g(z)\, dz_a \wedge d\bar{z}_{b \setminus \{\nu\}}$ is Γ_∞-invariant and we have

$$d\alpha = d(g(z)\, dz_a \wedge d\bar{z}_{b \setminus \{\nu\}}) = f(z)\, dz_a \wedge d\bar{z}_b = \omega$$

Since $f(\sigma_b(z))$ is a cusp form, f has analogous fourier series expansions with respect to arbitrary cusps k. Like above we obtain: If $\omega = f\, dz_a \wedge d\bar{z}_b$ is a canonical basis element of $H^n_{\text{cusp}}(\Gamma)$ of type (p,q) with $q \neq 0$, then we can choose the differential forms η_i satisfying $d\eta_i = \omega$ in sufficiently small neighbourhoods of the cusp classes k_i of type $(p, q-1)$. Therefore the representatives $\omega' \in W_0(K^n)$ of $[\omega] \in H^n_{\text{cusp}}(\Gamma)$ constructed in 7.2. are elements of $F^p(K^n)$, if we base our construction on the η_j's described above.

The case $b \neq \emptyset$ is to be treated somewhat differently. The canonical basis elements $\omega = f\, dz_1 \wedge \ldots \wedge dz_n$ with $f \in [\Gamma, (2, \ldots, 2)]_0$ admit an extension as holomorphic differential forms on $\overline{X}$. Thus $\omega \in W^0(K^n)$. (for a proof see Freitag [12], Satz 3.2). Since ω is of type $(n,0)$ we have $\omega \in F^n(K^n)$. A glance at the definition of the Hodge filtration on $H^m(\Gamma)$ shows:

7.4 Proposition.

$$\bigoplus_{\substack{p' \geq p \\ p' = n - \#b}} \bigoplus_{b \subset \{1, \ldots, n\}} [\Gamma^b, (2, \ldots, 2)]_0 \subset F^p(H^n(\Gamma))$$

By determining the weight filtration of the Eisenstein cohomology we shall show: $H^m_{\text{Eis}}(\Gamma) \cap W^0(H^m(\Gamma)) = \{0\}$. If we assume this result for the moment we have

$$Gr^W_0(H^m(\Gamma)) = H^m_{\text{squ}}(\Gamma)$$

This together with Propositions 3 and 4 yields by a symmetry argument $(h^{p,q}_m(\mathsf{H}^n/\Gamma) = h^{q,p}_m(\mathsf{H}^n/\Gamma))$:

7.5 Proposition. *The Hodge filtration on* $H^m_{\text{squ}}(\Gamma)$ *has the following form:*

$$F^p(H^m_{\text{squ}}(\Gamma)) = F^p(H^m_{\text{univ}}(\Gamma)) \oplus F^p(H^m_{\text{cusp}}(\Gamma)),$$

where $\quad F^p(H^m_{\text{univ}}(\Gamma)) = \begin{cases} H^m_{\text{univ}}(\Gamma) & \text{if } p \leq m/2 \\ \{0\} & \text{if } p > m/2 \end{cases}$

$$F^p(H^m_{\text{cusp}}(\Gamma)) = \begin{cases} \displaystyle\bigoplus_{\substack{p' \geq p \\ p' = n - \#b}} \bigoplus_{b \subset \{1, \ldots, n\}} [\Gamma^b, (2, \ldots, 2)]_0 & \text{if } m = n \\ \{0\} & \text{if } m \neq n \end{cases}$$

To determine the weight filtration of the Eisenstein cohomology $H^m_{\text{Eis}}(\Gamma)$, we investigate the geometrical and combinatorical structure of the Divisor $Y \subset \overline{X}$. Especially we consider the spectral sequences $S^{\cdot}$ (see Deligne [9], 3.2.8.1):

$$S^{\cdot} : \ldots \to H^{m-k-2}(\tilde{Y}^{k+1}, \varepsilon^{k+1}) \xrightarrow{d_1} H^{m-k}(\tilde{Y}^k, \varepsilon^k)$$
$$\xrightarrow{d_1} H^{m-k+2}(\tilde{Y}^{k-1}, \varepsilon^{k-1}) \to \ldots$$

The classical Hodge decompositions

$$H^l(\tilde{Y}^k, \varepsilon^k) = \bigoplus_{i+j=l} H^i(\tilde{Y}^k, \Omega^j_{\tilde{Y}^k}(\varepsilon^k))$$

define filtrations on $H^l(\tilde{Y}^k, \varepsilon^k)$. After suitable renumeration of these filtrations, $(S^{\cdot}, F)$ becomes a biregular bifiltered complex, such that the morphisms d_1 are strictly compatible with the filtration. Then $Gr^{\cdot}_F(H^{\cdot}(S^{\cdot})) \simeq H^{\cdot}(Gr^{\cdot}_F(S^{\cdot}))$ holds, i.e. the induced filtrations on the homology groups of $S^{\cdot}$ are determined by

$$F^p(H^{\cdot}(S^{\cdot})) = \bigoplus_{i \geq p} H^{\cdot}(Gr^i_F(S^{\cdot})) \,.$$

Deligne shows

$$H^{m-k}(S^{\cdot}) \simeq Gr^W_k(H^m(\Gamma K^{\cdot}))$$

and the filtration described above coincides with the Hodge filtration on $Gr^W_k(H^m(\Gamma K^{\cdot}))$. Therefore we contemplate the associated graded complexes $Gr^{\cdot}_F S^{\cdot}$:

$$\ldots \to H^{i-3}(\tilde{Y}^3, \Omega^{j-3}_{\tilde{Y}^3}(\varepsilon^3)) \xrightarrow{\partial} H^{i-2}(\tilde{Y}^2, \Omega^{j-2}_{\tilde{Y}^2}(\varepsilon^2))$$
$$\xrightarrow{\partial} H^{i-1}(\tilde{Y}^1, \Omega^{j-1}_{\tilde{Y}^1}(\varepsilon^1)) \xrightarrow{\partial} H^i(\overline{X}, \Omega^j_{\overline{X}}) \to 0$$

We shall show: For $0 \leq i \leq n$ we have

$$\dim Gr^n_F Gr^W_{n-i}(H^{n+i}(\Gamma))$$
$$= \dim \frac{\text{Kernel}\,(H^i(\tilde{Y}^{n-i}, \Omega^i_{\tilde{Y}^{n-i}}(\varepsilon^{n-i})) \xrightarrow{\partial} H^{i+1}(\tilde{Y}^{n-i-1}, \Omega^{i+1}_{\tilde{Y}^{n-i-1}}(\varepsilon^{n-i-1})))}{\text{Im}(H^{i-1}(\tilde{Y}^{n-i+1}, \Omega^{i-1}_{\tilde{Y}^{n-i+1}}(\varepsilon^{n-i+1})) \xrightarrow{\partial} H^i(\tilde{Y}^{n-i}, \Omega^i_{\tilde{Y}^{n-i}}(\varepsilon^{n-i})))}$$
$$= \dim H^{n+i}_{\text{Eis}}(\Gamma) \,.$$

Consequently $Gr^n_F Gr^W_{2n-m}(H^m(\Gamma)) = H^m_{\text{Eis}}(\Gamma)$ for $m \geq n$ and we obtain:

7.6 Proposition. *The weight filtration on $H^m(\Gamma)$ has the following form:*

$$W_{2n-m}(H^m(\Gamma)) = H^m_{\mathrm{squ}}(\Gamma) \oplus H^m_{\mathrm{Eis}}(\Gamma) = H^m(\Gamma)$$
$$W_{2n-m-1}(H^m(\Gamma)) = H^m_{\mathrm{squ}}(\Gamma)$$
$$\vdots$$
$$W_0(H^m(\Gamma)) = H^m_{\mathrm{squ}}(\Gamma)$$
$$W_{-1}(H^m(\Gamma)) = \{0\}$$

7.7 Proposition. *We have*

$$H^m_{\mathrm{Eis}}(\Gamma) \subset F^n(H^m(\Gamma)) \quad .$$

In order to prove the above propositions, we have to verify the statement about the homology groups of the complex $G^{\cdot}$

$$G^{\cdot} \quad \ldots \xrightarrow{\partial} H^i(\tilde{Y}^{n-i}, \Omega^i_{\tilde{Y}^{n-i}}(\varepsilon^{n-i})) \xrightarrow{\partial} \ldots \xrightarrow{\partial} H^{n-1}(\tilde{Y}^1, \Omega^{n-1}_{\tilde{Y}_1}(\varepsilon^1))$$
$$\xrightarrow{\partial} H^n(\overline{X}, \Omega^n_{\overline{X}}) \to 0 .$$

Since $\dim_{\mathbb{R}} \tilde{Y}^{n-j} = 2j$ for every j, we have

$$H^j(\tilde{Y}^{n-j}, \Omega^j_{\tilde{Y}_{n-j}}(\varepsilon^{n-j})) \simeq H^{2j}(\tilde{Y}^{n-j}, \varepsilon^{n-j}) ,$$

Thus $G^{\cdot}$ may be written as:

$$G^{\cdot} \quad \ldots \xrightarrow{\partial} H^{2i}(\tilde{Y}^{n-i}, \varepsilon^{n-i}) \xrightarrow{\partial} \ldots \xrightarrow{\partial} H^{2(n-1)}(\tilde{Y}^1, \varepsilon^1) \xrightarrow{\partial} H^{2n}(\overline{X}, \mathbb{C}) \to 0$$

Without loss of generality we may assume that Y is the union of smooth divisors. For let us suppose the contrary; then there exists a subgroup $\Gamma' \subset \Gamma$ of finite index in Γ, such that the divisor Y' compactifying $\mathbb{H}^n/\Gamma'$ is the union of smooth divisors (see appendix). If we consider the canonical projection $\pi : \mathbb{H}^n/\Gamma' \to \mathbb{H}^n/\Gamma$ and the induced mapping π^* on the cohomology groups

$$\pi^* : H^m(\mathbb{H}^n/\Gamma, \mathbb{C}) \to H^m(\mathbb{H}^n/\Gamma', \mathbb{C}) \quad ,$$

we observe that π^* is injective, respects the decomposition of the cohomology into Eisenstein- and square integrable part, and moreover induces a morphism of the respective mixed Hodge structures. Consequently

$$Gr^n_F Gr^W_{2n-m}(H^m(\Gamma)) = H^m_{\mathrm{Eis}}(\Gamma)$$

must hold, if only Γ' satisfies the analogous statement. Thus let us suppose Y is union of the smooth divisors $D_1, \ldots, D_r$. The choice of an order of these components of Y trivializes the sheaves ε^i, i.e. we obtain isomorphisms $\alpha^i : \varepsilon^i \simeq C$. The $\tilde{Y}^k$ split into disjoint connected divisors:

$$\tilde{Y}^k = \bigcup_{j \in J_k} \tilde{Y}^k_{(j)}$$

Especially $\tilde{Y}^1 = D_1 \,\dot{\cup}\, \ldots \,\dot{\cup}\, D_r$, i.e. $J_1 = \{1, \ldots, r\}$ and $\tilde{Y}^1_{(k)} = D_k$. Therefore we have the following decompositions:

$$H^{2i}(\tilde{Y}^{n-i}, \varepsilon^{n-i}) \simeq H^{2i}(\tilde{Y}^{n-i}, C) \simeq \bigoplus_{j \in J_{n-i}} H^{2i}(\tilde{Y}^{n-i}_{(j)}, C) \simeq \bigoplus_{j \in J_{n-i}} C$$

and $G^{\cdot}$ becomes

$$G^{\cdot} \quad \ldots \xrightarrow{\partial} \bigoplus_{j \in J_{n-i}} C \xrightarrow{\partial} \bigoplus_{j' \in J_{n-i-1}} C \xrightarrow{\partial} \bigoplus_{j'' \in J_{n-i-2}} C \xrightarrow{\partial} \ldots \xrightarrow{\partial} C \to 0$$

We can use $(\#J_\gamma) \times (\#J_{\gamma-1})$-matrices $(\partial_{\mu\nu})_{\mu \in J_\gamma, \nu \in J_{\gamma-1}}$ to describe the mappings ∂, where $\partial_{\mu\nu} \in C$ may be regarded as a linear mapping $\partial_{\mu\nu} : C \to C$. Later on we shall show: After suitable choice of the isomorphisms $H^{2i}(\tilde{Y}_{(j)}, C) \simeq C$ the $\partial_{\mu\nu}$ have the following form:

1) $\partial_{\mu\nu} = 0$, if $\tilde{Y}^{n-i}_{(\mu)} \not\subset \tilde{Y}^{n-i-1}_{(\nu)}$
2) let $\tilde{Y}^{n-i}_{(\mu)} \subset \tilde{Y}^{n-i-1}_{(\nu)}$ and let
   a) $i \leq n - 2$; then there are uniquely determined indices k_j,

   such that $\qquad \tilde{Y}^{n-i}_{(\mu)} = D_{k_1} \cap \ldots \cap D_{k_{n-i}}$

   and $\qquad\qquad \tilde{Y}^{n-i-1}_{(\nu)} = D_{k_1} \cap \ldots \cap \hat{D}_{k_\alpha} \cap \ldots \cap D_{k_{n-i}}$,

   since this is valid locally and by construction of Y arbitrary intersections of the divisors $D_1, \ldots, D_r$ are connected (see appendix). We have:

   $$\partial_{\mu\nu} = (-1)^\alpha$$

   b) $i = n - 1$; then $(\partial_{\mu\nu})$ is a $(r \times 1)$-matrix and we have

   $$(\partial_{\mu\nu}) = (1, \ldots, 1) .$$

The above remarks enable us to compute the homology groups of the complex $G^{\cdot}$. First let $i \leq n - 2$; in this case we only need to investigate a single cusp. If we take a look at the construction of Σ and Y in Ehlers [11] we can state (see appendix):

There is a 1:1-correspondence between the k-simplices of Σ/Λ and the components $\tilde{Y}^k_{(j)}$ of $\tilde{Y}^k$. Let $v_1, \ldots, v_r$ denote the 1-simplices associated to the components $D_1, \ldots, D_r$ of $\tilde{Y}$; then the simplex $\sigma_k \in \Sigma/\Lambda^{(k)}$ spanned by

$v_{i1}, \ldots, v_{ik}$ corresponds to the connected component $D_{i1} \cap \ldots \cap D_{ik}$ of $\tilde{Y}^k$; if $\langle v_{i1}, \ldots, v_{ik}\rangle \notin \Sigma/\Lambda^{(k)}$, then $D_{i1} \cap \ldots \cap D_{ik} = \emptyset$. If we compare the mappings ∂ of the complex $G^{\cdot}$ with the boundary operators ∂' of the simplicial complex Σ/Λ we observe: The complexes

$$
G^{\cdot} \quad \ldots \to \bigoplus_{j \in J_{n-i+1}} C \xrightarrow{\ \partial\ } \bigoplus_{j' \in J_{n-i}} C \xrightarrow{\ \partial\ } \bigoplus_{j'' \in J_{n-i-1}} C \to \ldots
$$

$$
\text{and} \quad \to \bigoplus_{k=1}^{h} \bigoplus_{\sigma_{i-1}} C \xrightarrow{\oplus_{1}^{h} \partial'} \bigoplus_{k=1}^{h} \bigoplus_{\sigma_{i}} C \xrightarrow{\oplus_{1}^{h} \partial'} \bigoplus_{k=1}^{h} \bigoplus_{\sigma_{i+1}} C \to \ldots
$$

are isomorphic. Hence

$$
H^{n-i}(G^{\cdot}) \simeq \bigoplus_{k=1}^{h} H_i(\Sigma/\Lambda, C) \quad \text{for} \quad 0 \le i \le n-2 \ .
$$

Remark. Σ/Λ is not a customary simplicial complex in the sense of algebraic topology, nevertheless the homology groups of the complex

$$
\ldots \to \bigoplus_{\sigma_{i-1}} C \xrightarrow{\ \partial'\ } \bigoplus_{\sigma_{i}} C \xrightarrow{\ \partial'\ } \bigoplus_{\sigma_{i+1}} C
$$

coincide with the singular homology groups $H_i(\Sigma/\Lambda, C)$. To see this, we only have to observe that the complexes Σ resp. Σ/Λ together with their boundary operators ∂' may be replaced by the customary simplicial complexes obtained by intersecting Σ resp. Σ/Λ with $A_1 := \{x \in \mathbf{R}_+^n \mid x_1 \cdots x_n = 1\}$. Since $H^{\cdot}(\Sigma) = 0$ we have $H_i(\Sigma/\Lambda, C) \simeq H^i(\Lambda)$. To compute the group cohomology $H^i(\Lambda)$ we recall that Λ acts on

$$
D := \{z \in H^n \mid Ny = 1\} .
$$

We have $H^{\cdot}(D) = 0$, thus

$$
H^i(\Lambda) \simeq H^i(D/\Lambda, C) \simeq H^i((S^1)^{n-1}, C),
$$

since $D/\Lambda \simeq (S^1)^{n-1}$ (see Chap. III, §2). Hence:

$$
\dim H^{n-i}(G^{\cdot}) = h \cdot \dim H^i(\Lambda) = h \cdot \dim H^i((S^1)^{n-1}, C)
$$

$$
= h \binom{n-1}{i} = \dim H_{\text{Eis}}^{n+i}
$$

for $0 \le i \le n-2$. We have to add the treatment of the remaining case $i = n-1$. If we cut off the complex $G^{\cdot}$ at

$$
\bigoplus_{j \in J_1} C \xrightarrow{\ \partial\ } C
$$

we obtain the complex

$$\tilde{G}^{\cdot} : \quad \ldots \xrightarrow{\partial} \bigoplus_{j \in J_1} \mathbf{C} \xrightarrow{\vdots} 0 \to 0$$

Analogous to the case $0 \le i \le n-2$ we can show:

$$H^1(\tilde{G}^{\cdot}) := H(\xrightarrow{\partial} \bigoplus_{j \in J_1} \mathbf{C} \to 0) \simeq \bigoplus_{j=1}^{h} H^{n-1}(\Lambda)$$

Therefore we have:

$$\dim(\mathrm{Kernel}(\bigoplus_{j \in J_1} \mathbf{C} \to 0)) = \dim \bigoplus_{j \in J_1} \mathbf{C} = \#J_1$$

and

$$\dim(\mathrm{Im}(\bigoplus_{j' \in J_2} \mathbf{C} \xrightarrow{\partial} \bigoplus_{j \in J_1} \mathbf{C})) = \#J_1 - \dim \bigoplus_{j=1}^{h} H^{n-1}(\Lambda) = \#J_1 - h \quad .$$

Considering the complex

$$G^{\cdot} : \quad \ldots \to \bigoplus_{j' \in J_2} \mathbf{C} \xrightarrow{\partial} \bigoplus_{j \in J_1} \mathbf{C} \xrightarrow{\partial} \mathbf{C} \to 0$$

we obtain

$$\dim(\mathrm{Kernel}(\bigoplus_{j \in J_1} \mathbf{C} \xrightarrow{\partial} \mathbf{C})) = \#J_1 - 1$$

since $\partial : \bigoplus_{j \in J_1} \mathbf{C} \to \mathbf{C}$ is surjective. We already know:

$$\dim(\mathrm{Im}(\bigoplus_{j' \in J_2} \mathbf{C} \xrightarrow{\partial} \bigoplus_{j \in J_1} \mathbf{C})) = \#J_1 - h$$

Therefore we have:

$$\dim H^1(G^{\cdot}) = (\#J_1 - h) - (\#J_1 - 1) = h - 1 = \dim H_{\mathrm{Eis}}^{2n-1}(\Gamma) \quad .$$

To complete our proof we have to verify that the mappings $\partial_{\mu\nu} : \mathbf{C} \to \mathbf{C}$ resp.

$$\partial_{\mu\nu} : H^{2i}(\tilde{Y}_{(\mu)}^{n-i}, \mathbf{C}) \to H^{2(i+1)}(\tilde{Y}_{(\nu)}^{n-i-1}, \mathbf{C})$$

actually have the form described above. We define the isomorphisms α : $H^{2i}(\tilde{Y}_{(j)}^{n-i}, \mathbb{C}) \simeq \mathbb{C}$ by

$$\alpha([\omega]) := (2\pi i)^{n-i} \int_{\tilde{Y}_{(j)}^{n-i}} \omega \quad .$$

Thus it remains to verify the following staement: Let $[\omega] \in H^{2i}(\tilde{Y}_{(\mu)}^{n-i}, \mathbb{C})$ and let $[\omega'] = \partial_{\mu\nu}([\omega]) \in H^{2(i+1)}(\tilde{Y}_{(\nu)}^{n-i-1}, \mathbb{C})$, then:

$$\alpha([\omega']) = \begin{cases} \pm\alpha([\omega]) & \text{if } \tilde{Y}_{(\mu)}^{n-i} \subset \tilde{Y}_{(\nu)}^{n-i-1}, \quad \begin{matrix} \text{where the sign is to be} \\ \text{chosen as above} \end{matrix} \\ 0 & \text{if } \tilde{Y}_{(\mu)}^{n-i} \not\subset \tilde{Y}_{(\nu)}^{n-i-1} \end{cases}$$

Let ω be a differential form on $\tilde{Y}_{(\mu)}^{n-i}$, then $\partial_{\mu\nu}([\omega]) = \partial([\omega])|\tilde{Y}_{(\nu)}$. We shall show that starting with ω, the corresponding representative ω' of $\partial([\omega])$ has the required properties. The reader who looks carefully at the definition of the Poincaré-residues and at the construction of the spectral sequences $S^{\cdot}$ (see Deligne [1], 3.2.8.1) will establish that ω' may be constructed in the following way:

1) Extend ω to ω^{ext} on $\overline{X}$.
2) For $i = 1, \ldots, r$ choose C^{∞}-functions t_i with the following property: Let $z \in D_i$, then there exists an open neighbourhood $U(z) \subset \overline{X}$ of z, such that $t_i = 0$ is a local defining equation of D_i in $U(z)$. (The functions t_i may be constructed via partition of unity out of local defining equations of D_i).
3) Take $\omega^{ext} \wedge \dfrac{dt_{i1}}{t_{i1}} \wedge \ldots \wedge \dfrac{dt_{ik}}{t_{ik}}$, if $\tilde{Y}_{(\mu)}^{n-i} = D_{i1} \cap \ldots \cap D_{ik}$.
   This differential form is defined in an open neighbourhood U of $\tilde{Y}_{(\mu)}^{n-i}$. Multiplication with a suitable C^{∞}-function ϕ satisfying $\phi \equiv 1$ in an open neighbourhood $V \subset\subset U$ of $\tilde{Y}_{(\mu)}^{n-i}$ and $\phi \equiv 0$ outside U yields a globally defined differential form

$$\phi \cdot \omega^{ext} \wedge \frac{dt_{i1}}{t_{i1}} \wedge \ldots \wedge \frac{dt_{ik}}{t_{ik}}$$

4) Take $d(\phi \cdot \omega^{ext} \wedge \frac{dt_{i1}}{t_{i1}} \wedge \ldots \wedge \frac{dt_{ik}}{t_{ik}})$
5) Let $\tilde{Y}_{(\nu)}^{n-i-1} = D_{i1} \cap \ldots \cap \widehat{D_{i\alpha}} \cap \ldots \cap D_{ik}$. Then we have:

$$\omega'|\tilde{Y}_{(\nu)}^{n-i-1} = (-1)^{\alpha} d(\phi \cdot \omega^{ext} \wedge \frac{dt_{i\alpha}}{t_{i\alpha}})$$

(that's the Poincaré-residue). We have to compute:

$$\alpha([\omega']) = (2\pi i)^{n-i-1} \int_{\tilde{Y}_{(\nu)}^{n-i-1}} (-1)^{\alpha} d(\phi \cdot \omega^{\text{ext}} \wedge \frac{dt_{i\alpha}}{t_{i\alpha}})$$

$$= (-1)^{\alpha}(2\pi i)^{n-i-1} \lim_{\epsilon \to 0} \int_{\tilde{Y}_{(\mu)}^{n-i} \times S_{\epsilon}} \phi \cdot \omega^{\text{ext}} \wedge \frac{dt_{i\alpha}}{t_{i\alpha}},$$

where $S_{\epsilon} := \{t_{i\alpha} | |t_{i\alpha}| = \epsilon\}$

$$= (-1)^{\alpha}(2\pi i)^{n-i-1} \int_{\tilde{Y}_{(\mu)}^{n-i}} \omega^{\text{ext}}(t_{i\alpha} = 0) \cdot \lim_{\epsilon \to 0} \int_{S_{\epsilon}} \frac{dt_{i\alpha}}{t_{i\alpha}}$$

$$= (-1)^{\alpha}(2\pi i)^{n-i} \int_{\tilde{Y}_{(\mu)}^{n-i}} \omega = (-1)^{\alpha} \alpha([\omega])$$

Moreover we have $\omega'/\tilde{Y}_{(\nu)}^{n-i-1} = 0$, if $\tilde{Y}_{(\mu)}^{n-i} \not\subset \tilde{Y}_{(\nu)}^{n-i-1}$. Now the proof is complete. □

The Hodge Numbers of H^n/Γ. The Propositions 1 through 7 contain the whole information about the weight- and Hodge filtration on $H^m(\Gamma)$, i.e. we know the mixed Hodge structure $(H^m(\Gamma), F, W)$, which is equivalent to the Hodge decomposition

$$H^m(\Gamma) = \bigoplus_{p,q} H_m^{p,q}(\Gamma) \quad ,$$

where $H_m^{p,q}(\Gamma)$ is defined as

$$H_m^{p,q}(\Gamma) := Gr_F^p Gr_{\bar{F}}^q Gr_{p+q}^{W[m]}(H^m(\Gamma))$$

Summarizing our results we obtain:

7.8 Theorem. *We have*

1) $H_0^{0,0}(\Gamma) = H^0(\Gamma)$

2) $H_m^{p,q}(\Gamma) = {}_{\text{univ}} H_m^{p,q}(\Gamma) \oplus {}_{\text{cusp}} H_m^{p,q}(\Gamma) \oplus {}_{\text{Eis}} H_m^{p,q}(\Gamma)$

where

$${}_{\text{univ}} H_m^{m/2,m/2}(\Gamma) = H_{\text{univ}}^m(\Gamma); \quad {}_{\text{univ}} H_m^{p,q}(\Gamma) = \{0\} \text{ otherwise}$$

$${}_{\text{cusp}} H_n^{p,q}(\Gamma) = \bigoplus_{\substack{b \subset \{1,\ldots,n\} \\ p=n-\#b,\, q=\#b}} [\Gamma^b, (2,\ldots,2)]_0;$$

$${}_{\text{cusp}} H_m^{p,q}(\Gamma) = \{0\} \text{ otherwise}$$

$${}_{\text{Eis}} H_m^{n,n}(\Gamma) = H_{\text{Eis}}^m(\Gamma); \quad {}_{\text{Eis}} H_m^{p,q}(\Gamma) = \{0\} \text{ otherwise}$$

We may represent this result by considering the following Hodge diagram:

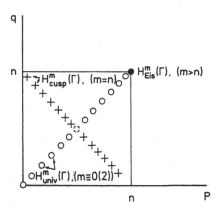

The Hodge numbers $h_m^{p,q} = h_m^{p,q}(\mathbf{H}^n/\Gamma)$ are defined as

$$h_m^{p,q} := \dim H_m^{p,q}(\Gamma)$$

Therefore we obtain from Theorem 1:

7.9 Theorem. *The Hodge numbers of the Hilbert modular variety are given by:*

a) $m = 0$: $h_0^{0,0} = 1$
b) $m = 2n$: $h_m^{p,q} = 0$
c) $0 < m < 2n$: $h_m^{p,q} = {}_{\mathrm{univ}}h_m^{p,q} + {}_{\mathrm{cusp}}h_m^{p,q} + {}_{\mathrm{Eis}}h_m^{p,q}$

with:

$$_{\mathrm{univ}}h_m^{m/2,m/2} = \binom{n}{m/2}; \quad {}_{\mathrm{univ}}h_m^{p,q} = 0 \quad \text{otherwise}$$

$$_{\mathrm{cusp}}h_n^{p,q} = h_{\mathrm{cusp}}^{p,q} \text{ for } p + q = n, \quad {}_{\mathrm{cusp}}h_m^{p,q} = 0 \quad \text{otherwise}$$

$$\text{where} \quad h_{\mathrm{cusp}}^{p,q} := \sum_{\substack{b \subset \{1,\dots,n\} \\ p=n-\#b, q=\#b}} \dim[\Gamma^b, (2,\dots,2)]_0$$

$$_{\mathrm{Eis}}h_m^{n,n} = \begin{cases} h\binom{n-1}{m-n} & \text{if } n < m < 2n - 1 \\ h - 1 & \text{if } m = 2n - 1 \mathrm{cr} \end{cases} \qquad {}_{\mathrm{Eis}}h_m^{p,q} = 0 \quad \text{otherwise}$$

Appendix

We give a short summary of the basic definitions and results of Ehlers [11] used in our context.

1) Let $t \subset \mathbf{R}^n$ be a free $\mathbf{Z}$-module of rank n. Let $v_1, \ldots, v_p \in t$ be linearly independent. $(v_1, \ldots, v_p)$ is called p-simplex, if $t/\langle v_1, \ldots, v_p \rangle_{\mathbf{Z}}$ has no elements of finite order. To each p-simplex $(v_1, \ldots, v_p)$ we associate the set

$$\sigma = \langle v_1, \ldots, v_p \rangle_{\mathbf{R}_+} := \{\textstyle\sum t_i v_i \mid t_i \geq 0\}$$

Also we often call σ a p-simplex. The $\langle v_{i1}, \ldots, v_{ik} \rangle_{\mathbf{R}_+}$, $1 \leq i_j \leq p$ are called k-faces of σ.

2) A set Σ of simplices is called complex, if the following properties hold:

(i) $\sigma, \sigma' \in \Sigma \Rightarrow int(\sigma) \cap int(\sigma') = \emptyset$ for $\sigma \neq \sigma'$, $\sigma \cap \sigma' \in \Sigma$,

where $int(\sigma) = \{\sum t_i v_i \mid t_i > 0\}$

(ii) $\tau \in \Sigma \Rightarrow St(\tau) := \{\sigma \in \Sigma \mid \tau \text{ face of } \sigma\}$ is a finite set.

(iii) $\tau \in \Sigma$, $\dim \tau < n \Rightarrow \tau$ is face of a suitable n-simplex $\sigma \in \Sigma$.

3) To each simplicial complex Σ we can associate a complex manifold X_Σ in the following way: Let $\sigma, \sigma' \in \Sigma^{(n)} := \{n\text{-simplices of } \Sigma\}$ and $G_{\sigma'\sigma} \in Gl_n\mathbf{Z}$ the matrix defined by $\sigma'^t = G_{\sigma'\sigma} \cdot \sigma$; for arbitrary $A \in Gl_n\mathbf{Z}$ and $z = (z_1, \ldots, z_n) \in \mathbf{C}^n$ let

$$z^A := \begin{pmatrix} z_1^{a_{11}} & \cdots & z_n^{a_{1n}} \\ & \vdots & \\ z_1^{a_{n1}} & \cdots & z_n^{a_{nn}} \end{pmatrix} \in \mathbf{C}^n$$

Then we define

$$X_\Sigma := \bigcup_{\sigma \in \Sigma^{(n)}} (\mathbf{C}^n)_\sigma / \sim$$

where the glueing relation is given by the mappings

$$g_{\sigma'\sigma} : (\mathbf{C}^n)_\sigma \rightarrow (\mathbf{C}^n)_{\sigma'}$$
$$z \rightarrow z^{G^*_{\sigma'\sigma}}$$

where $G^*_{\sigma'\sigma} := (G^{-1}_{\sigma'\sigma})^t$.

4) To each $\tau = \langle v_1, \ldots, v_r \rangle \in \Sigma^{(r)}$ there is associated a connected submanifold $F_\tau \subset X_\Sigma$ of codimension r. In coordinates we have $F_\tau \cap (\mathbf{C}^n)_\sigma = \{z \mid z_1 = \ldots = z_r = 0\}$, if $\sigma = \langle v_1, \ldots, v_r, v_{r+1}, \ldots, v_n \rangle$.

5) $D := \bigcup_{\dim \tau = 1} F_\tau$ is a divisor with normal crossings in X_Σ. For $\tau = \langle v_1, \ldots, v_r \rangle \in \Sigma^{(r)}$ we have:

$$F_\tau = F_{v_1} \cap \ldots \cap F_{v_r} \quad .$$

But if $(v_1, \ldots, v_r) \notin \Sigma^{(r)}$, then : $F_{v_1} \cap \ldots \cap F_{v_r} = \emptyset$.

6) Let $\Gamma \subset \Gamma_K$ be a congruence subgroup of some Hilbert modular group Γ_K; let t be the translation lattice and Λ the group of multipliers of Γ. Then there exists a simplicial complex Σ having the following properties:

(i) $|\Sigma| := \bigcup_{\sigma \in \Sigma} \sigma = (\mathbf{R}_+^n) \cup \{0\}$

(ii) Λ acts on Σ, i.e. $\sigma \in \Sigma$, $\varepsilon \in \Lambda \Rightarrow \varepsilon\sigma \in \Sigma$

(iii) $\sigma \in \Sigma$, $\varepsilon \in \Lambda$, $\varepsilon \neq 1 \Rightarrow \dim(\sigma \cap \varepsilon\sigma) \leq 1$

When replacing Γ by a suitable subgroup $\Gamma' \subset \Gamma$ of finite index, we can attain $\dim(\sigma \cap \varepsilon\sigma) = 0$ in (iii).

7) There is an open neighbourhood $U(D) \subset X_\Sigma$ of the divisor D and a biholomorphic mapping π:

$$\pi: \quad U(D)\backslash D \simeq U_C/t$$

where
$$U_C := \{z \in \mathbf{H}^n \mid Ny > C > 0\}$$

8) Λ acts discontinously and freely on $U(D)$, such that:

(i) $x \in D$, $\varepsilon \in \Lambda \Rightarrow \varepsilon x \in D$, i.e. Λ acts on D.

(ii) $\pi : U(D)\backslash D \simeq U_C/t$ is equivariant with respect to the natural action of Λ on U_C/A.

9) Consequently $U(D)/\Lambda$ is a complex manifold, $Y_\infty := D/\Lambda$ a divisor with normal crossings in $U(D)/\Lambda$. We have:

$$(U(D)\backslash D)/\Lambda \simeq (U(D)/\Lambda)\backslash Y_\infty \simeq U_C/\Gamma_\infty \subset \mathbf{H}^n/\Gamma$$

is an open neighbourhood of the cusp $\{\infty\}$. We define a mapping

$$f : U(D)/\Lambda \to U_C/\Gamma_\infty \cup \{\infty\} \subset X_\Gamma$$

by $f(Y_\infty) := \{\infty\}$ and by defining $f|(U(D)/\Lambda)\backslash Y_\infty$ as the induced isomorphism $\pi^* : (U(D)/\Lambda)\backslash Y_\infty \simeq U_C/\Gamma_\infty$. Then f yields a resolution of the singularity $\{\infty\}$ of X_Γ with the properties described in the introduction.

10) The action of Λ on $U(D)$ is such that each $\varepsilon \in \Lambda$ gives a rise to a biholomorphic mapping

$$\varepsilon : F_\sigma \simeq F_{\varepsilon\sigma} \quad .$$

If $\Gamma' \in \Gamma$ is chosen as in 6. with $\sigma \cap \varepsilon\sigma = \{0\}$ for $\varepsilon \neq 1$, then $F_\sigma \cap F_{\varepsilon\sigma} = \emptyset$ for $\varepsilon \neq 1$. To see this let $\sigma = (v_1, \ldots, v_r)$ and $\varepsilon\sigma = (v_1', \ldots, v_r')$ and $\tau = (v_1, \ldots, v_r, v_1', \ldots, v_r')$; then τ can not be a simplex of $\Sigma^{(2r)}$, since that would contradict $\tau \cap \varepsilon\tau = \{0\}$. Thus:

$$F_\sigma \cap F_{\varepsilon\sigma} = F_{v_1} \cap \ldots \cap F_{v_r} \cap F_{v_1'} \cap \ldots \cap F_{v_r'} = \emptyset$$

Therefore we have

$$F_\sigma/\Lambda \simeq F_\sigma$$

for each $\sigma \in \Sigma$. Especially $F_{v_i}/\Lambda \simeq F_{v_i}$ for each 1-simplex v_i. Consequently $Y_\infty = D/\Lambda$ is the union of smooth divisors $D_1, \ldots, D_r$; moreover arbitrary intersections of these divisors are connected, since for suitable $v_{i1}, \ldots, v_{ik}$ we have:

$$D_{i1} \cap \ldots \cap D_{ik} \simeq F_{v_{i1}} \cap \ldots \cap F_{v_{ik}} \simeq F_\tau$$

if $\tau = \langle v_{i1}, \ldots, v_{ik} \rangle \in \Sigma^{(k)}$ and

$$D_{i1} \cap \ldots \cap D_{ik} = \emptyset$$

if $\tau \notin \Sigma^{(k)}$. Obviously we have a 1:1 correspondence between the 1-simplices $v_1, \ldots, v_r$ of Σ/Λ and the smooth divisors $D_1, \ldots, D_r$. The simplices $\langle v_{i1}, \ldots, v_{ik} \rangle_{\mathbf{R}_+}$ correspond to the submanifolds of codimension k given by $D_{i1} \cap \ldots \cap D_{ik}$, which coincide with the connected components $\tilde{Y}^k_{(j)}$ of $\tilde{Y}^k$ (see above).

Appendices

I. Algebraic Numbers

We give a brief introduction, without proofs, to the theory of algebraic numbers.

A complex number is called **algebraic** if it is a root of a non-vanishing polynomial with rational coefficients

$$a_n a^n + \ldots + a_1 a + a_0 = 0, \quad a_j \in \mathbf{Q} \text{ for } 0 \leq j \leq n, \quad a_n \neq 0.$$

We may assume that the polynomial is **monic**, i.e. $a_n = 1$.

If a is a root of a monic polynomial whose coefficients are rational integers we call a an **(algebraic) integer**.

A monic polynomial with rational coefficients of minimal degree

$$P \in \mathbf{Q}[x] \quad , \quad P(a) = 0$$

is called **minimal polynomial** of a.

AI.1. *The minimal polynomial of an algebraic number is uniquely determined. It is separable, i.e. it has no multiple root. The minimal polynomial of an algebraic integer has (rational) integral coefficients.*

Notation. The degree of an algebraic number is the degree of the minimal polynomial. The different roots of the minimal polynomial are called **conjugates** of an algebraic number a. We often denote them by

$$a^{(1)}, \ldots, a^{(n)} \quad (a \text{ is one of them}).$$

AI.2. *The set $\overline{\mathbf{Q}}$ of all algebraic numbers is a subfield of $\mathbf{C}$ which contains $\mathbf{Q}$. The set of all algebraic integers $\overline{\mathbf{Z}}$ is a subring of $\overline{\mathbf{Q}}$ with the property*

$$\overline{\mathbf{Z}} \cap \mathbf{Q} = \mathbf{Z}.$$

AI.3. *Let P be a monic polynomial with algebraic coefficients $(P \in \overline{\mathbf{Q}}[x])$. The roots of P are algebraic. If the coefficients are integral $(P \in \overline{\mathbf{Z}}[x])$, the roots are algebraic integers.*

Number Fields. Let K be a subfield of the field of complex numbers ($\mathbf{Q} \subset K \subset \mathbf{C}$). We may consider K as a vector space over $\mathbf{Q}$. K is called an **algebraic number field** if the dimension of this vector space is finite. This dimension is called the degree of K and denoted by

$$n = [K : \mathbf{Q}] = \dim_{\mathbf{Q}} K \,.$$

(One can always define the degree of an arbitrary field K with respect to a subfield k. $[K : k] = \dim_k K \leq \infty$.) The elements of algebraic number fields are always algebraic numbers and each algebraic number is contained in some algebraic number field K. The smallest K which contains a is denoted by

$$K = \mathbf{Q}(a) \,.$$

Conjugate Fields. Let K be an algebraic number field of degree n. An imbedding of K into the field of complex numbers is a mapping

$$\varphi : K \longrightarrow \mathbf{C}$$

with the properties

1) $$\varphi(a) = a \quad \text{for } a \in \mathbf{Q} \,,$$

2) $$\varphi(a \dotplus b) = \varphi(a) \dotplus \varphi(b) \,.$$

The image of K is a subfield of $\mathbf{C}$ isomorphic to K, $K \overset{\sim}{\longrightarrow} \varphi(K)$.

AI.4. *An algebraic number field K of degree n admits precisely n different imbeddings into the field of complex numbers.*

We usually arrange the imbeddings in a certain order and denote them by

$$K \longrightarrow K^{(j)} \subset \mathbf{C} \,,$$
$$a \longmapsto a^{(j)} \,, \quad j = 1, \ldots, n \,.$$

AI.5. *If a is an element of K, the images under the imbeddings are conjugates of a. Each conjugate occurs under these images with multiplicity*

$$[K : \mathbf{Q}(a)] = [K : \mathbf{Q}]/degree(a) \,.$$

(We have $degree(a) = [\mathbf{Q}(a) : \mathbf{Q}]$.)

Notation. Consider the n different imbeddings

$$K \longrightarrow K^{(j)} \quad, \quad 1 \leq j \leq n\,.$$

trace:

$$Sa = S_{K/\mathbf{Q}}(a) = a^{(1)} + \ldots + a^{(n)}\,,$$

norm:

$$Na = N_{K/\mathbf{Q}}(a) = a^{(1)} \cdot \ldots \cdot a^{(n)}\,.$$

AI.6. *Trace and norm of an element $a \in K$ are rational numbers. If a is an algebraic integer they are rational integers.*

The Discriminant. Let $a_1, \ldots, a_n$ be elements of our algebraic number field K (of degree n). The discriminant of $(a_1, \ldots, a_n)$ is defined by

$$d(a_1, \ldots, a_n) = (\det A)^2\,,$$

where

$$A = \begin{pmatrix} a_1^{(1)} & \cdots & a_1^{(n)} \\ \vdots & & \vdots \\ a_n^{(1)} & \cdots & a_n^{(n)} \end{pmatrix}.$$

Obviously

$$A \cdot A' = (S(a_i a_j))_{1 \leq i,j \leq n}\,.$$

We obtain that $d(a_1, \ldots, a_n)$ is a rational number (which does not depend on the ordering of the imbeddings). If all $a_1, \ldots, a_n$ are algebraic integers, the discriminant is a rational integer.

AI.7. *The discriminant $d(a_1, \ldots, a_n)$ is 0 iff the elements $a_1, \ldots, a_n$ are linearly dependent over $\mathbf{Q}$. We moreover have*

$$d(a_1, \ldots, a_n) = d(b_1, \ldots, b_n) \quad \text{if} \quad \mathbf{Z}a_1 + \ldots + \mathbf{Z}a_n = \mathbf{Z}b_1 + \ldots + \mathbf{Z}b_n\,.$$

A lattice $\mathbf{m}$ in K is an additive subgroup of K which generates K as $\mathbf{Q}$-vector space and which is isomorphic to the group $\mathbf{Z}^n$. This means that we can find a $\mathbf{Q}$-basis $a_1, \ldots, a_n$ of K with

$$\mathbf{m} = \mathbf{Z}a_1 + \ldots + \mathbf{Z}a_n\,.$$

By AI.7 the discriminant

$$d(\mathbf{m}) := d(a_1, \ldots, a_n)$$

of a lattice is a well defined rational number, different from 0.

The Ring of Integers. We denote by

$$\mathrm{o} = \mathrm{o}_K = K \cap \bar{\mathbf{Z}}$$

the ring of algebraic integers in our number field K (of degree n).

AI.8. o *is a lattice in* K, *i.e.*

a) $K = \mathbf{Q} \cdot \mathrm{o}$,

b) $\mathrm{o} = \mathbf{Z}a_1 + \ldots + \mathbf{Z}a_n$.

The **discriminant** of K is by definition the discriminant of the lattice o, i.e.

$$d_K := d(\mathrm{o}) \qquad (= d(a_1, \ldots, a_n)) \, .$$

(The discriminant is a very important invariant of K. It can be shown that always $d \equiv 0$ or $1 \bmod 4$ and that for given d there exist only finitely many number fields with discriminant d, especially: $n \to \infty \Rightarrow d \to \infty$.)

Units. The invertible elements of o are the so-called units of K

$$\mathrm{o}^* = \{\varepsilon \in \mathrm{o} \mid \varepsilon \neq 0 \, , \, \varepsilon^{-1} \in \mathrm{o}\} \, .$$

Special units of K are the roots of unity which are contained in K

$$W = \{\zeta \in K \mid \zeta^h = 1 \text{ for some } h \in \mathbf{N}\} \, .$$

Of course o^* is a multiplicative group and W is a subgroup.

Before we describe the structure of o^* we have to define a further invariant r of K:

An imbedding φ of K is called **real** if the image $\varphi(K)$ is contained in the field $\mathbf{R}$ of real numbers. We denote by r_1 the number of real imbeddings of K. The number of non-real imbeddings is even, because $a \mapsto \overline{\varphi(a)}$ is also an imbedding of K. We may therefore write

$n = r_1 + 2r_2$

$r_1 = $ number of real imbeddings

$r_2 = $ number of pairs $(\varphi, \bar{\varphi})$ of complex conjugate non-real imbeddings.

We define the invariant r by

$$r = r_1 + r_2 - 1 \qquad (\leq n - 1) \, .$$

A famous theorem of Dirichlet states

AI.9. *The group of roots of unity in an algebraic number field is a finite cyclic group. The structure of the whole unit group is given by*

$$o^* \cong W \times Z^r .$$

Special case: The field K is called totally real if it admits only real imbeddings. In this case we have $r = n - 1$ and, of course, $W = \{1, -1\}$.

It is often useful to select one representative from each pair $\{\varphi, \overline{\varphi}\}$ of complex conjugate non-real imbeddings and to arrange the imbeddings as follows:

$$K \longrightarrow R^{r_1} \times C^{r_2}$$

$$a \longmapsto (a^{(1)}, \ldots, a^{(r_1+r_2)}) .$$

Here

$$a \longmapsto a^{(j)} \quad , \quad j = 1, \ldots, r_1$$

denotes the real imbeddings and

$$a \longmapsto a^{(j)} \quad , \quad r_1 < j \le r_1 + r_2$$

the set of representatives of non-real imbeddings. The following formulation of the unit theorem is roughly equivalent with AI.9 (but more on the line of a proof).

AI.10. *An integer $a \in o$ is a unit if and only if*

$$| Na | = | a^{(1)} \cdot \ldots \cdot a^{(n)} | = 1 .$$

It is a root of unity if and only if

$$| a^{(1)} | = \ldots = | a^{(n)} | = 1 .$$

The mapping

$$a \longmapsto (\log | a^{(1)} |, \ldots, \log | a^{(r_1+r_2)} |)$$

defines an isomorphism of o^*/W to a sublattice of the vector space

$$\{x \in R^{r_1+r_2} \mid x_1 + \ldots + x_{r_1+r_2} = 0\} .$$

Ideals. A subset $a \subset K, a \ne \{0\}$ of an algebraic number field is called an **ideal** if the following conditions hold:

a) a is an additive group.
b) $a \in a, b \in o \Rightarrow a \cdot b \in a$.
c) There exists a non-zero element $a \in o$ with $a \cdot a \subset o$.

AI.11. *Each ideal* **a** *is a lattice in* K, *i.e.*

$$\mathbf{a} = Za_1 + \ldots + Za_n \,,$$

where $a_1, \ldots, a_n$ *is a* **Q**-*basis of* K. *We especially can define the discriminant* $d(\mathbf{a})$ *of an ideal.*

Product of Ideals. If $a_1, \ldots, a_m$ is a finite system of numbers in K, not all of them 0, then

$$(a_1, \ldots, a_m) = \left\{ \sum_{j=1}^{m} b_j a_j \mid b_j \in \mathbf{o} \right\}$$

is an ideal. An ideal is called **principal** if it can be generated by one element

$$\mathbf{a} = (a) \,.$$

The product of two ideals $\mathbf{a}, \mathbf{b}$ is defined by

$$\mathbf{a} \cdot \mathbf{b} = \left\{ \sum_{\text{finite number}} a_j b_j \mid a_j \in \mathbf{a}, b_j \in \mathbf{b} \right\}.$$

It is again an ideal and the formulae

$$(a)\mathbf{a} = a\mathbf{a} = \{ax \mid x \in \mathbf{a}\}$$
$$(a) \cdot (b) = (ab)$$

and more generally

$$(a_1, \ldots, a_m)(b_1, \ldots, b_l) = (a_i b_j)$$

hold.

From AI.11 it follows that each ideal is finitely generated. But one can show even more:

AI.12. *Let* a *be a non-zero element of an ideal* **a**. *There exists a second element* b *with*

$$\mathbf{a} = (a, b) \,.$$

AI.13. *The set* $\mathcal{I}$ *of all ideals of* K *is a group under the multiplication introduced above. The unit element of this group is the ideal* $\mathbf{o} = (1)$. *The inverse of an ideal* **a** *is given by*

$$\mathbf{a}^{-1} = \{x \in K \mid x\mathbf{a} \subset \mathbf{o}\} \,.$$

The set $\mathcal{H}$ of principal ideals is a subgroup of $\mathcal{I}$. The factor group

$$\mathcal{I}/\mathcal{H}$$

is a finite (abelian) group.

The order $h = \#\mathcal{I}/\mathcal{H}$ is a very important invariant of K, the so-called **class number** of K. The elements of $\mathcal{I}/\mathcal{H}$ are called **ideal classes**.

Prime Ideals. An ideal $\mathbf{p}$ of K is called **prime ideal** if it is integral ($\mathbf{p} \subset \mathbf{o}$) and if it satisfies

$$a \cdot b \in \mathbf{p}, \, a, b \in \mathbf{o} \implies a \in \mathbf{p} \quad \text{or} \quad b \in \mathbf{p}$$

(i.e., $\mathbf{o}/\mathbf{p}$ is an integral domain).

A fundamental result of algebraic number theory states

AI.14. *Each ideal (of our number field K) can be written in the form*

$$\mathbf{a} = \mathbf{p}_1^{j_1} \cdot \ldots \cdot \mathbf{p}_m^{j_m} \quad , \quad j_\nu \in \mathbb{Z} \text{ for } 1 \leq \nu \leq m.$$

Here $\mathbf{p}_1, \ldots, \mathbf{p}_m$ are pairwise distinct prime ideals. This representation is unique up to permutation of the factors. $\mathbf{a}$ is integral iff all the exponents j_ν are non-negative.

We note some immediate consequences of AI.13, AI.14.

1) An integral ideal $\mathbf{p} \subset \mathbf{a}$ is a prime ideal if and only if it is maximal:

$$\mathbf{p} \subset \mathbf{a} \subset \mathbf{o} \implies \mathbf{a} = \mathbf{p} \text{ or } \mathbf{a} = \mathbf{o},$$

or equivalently: $\mathbf{o}/\mathbf{p}$ is a field.

2) For two ideals $\mathbf{a}, \mathbf{b}$ the following two conditions are equivalent:

i)
$$\mathbf{a} \supset \mathbf{b},$$

ii)
$$\mathbf{a}^{-1}\mathbf{b} \text{ is integral}.$$

In this case we say that $\mathbf{a}$ divides $\mathbf{b}$

$$\mathbf{a} \mid \mathbf{b}.$$

Notice: In the case of principal ideals $\mathbf{a} = (a), \mathbf{b} = (b)$ we have the usual notion of divisibility

$$(a) \mid (b) \iff a^{-1}b \in \mathbf{o}.$$

The Norm of Ideals.

AI.15. *There is a unique mapping (the norm of ideals)*

$$\mathcal{N} : \mathcal{I} \longrightarrow \mathbf{Q} - \{0\}$$

with the following properties

1) $\mathcal{N}(\mathbf{a} \cdot \mathbf{b}) = \mathcal{N}(\mathbf{a}) \cdot \mathcal{N}(\mathbf{b})$.

If $\mathbf{a} \subset \mathbf{o}$ *is integral*

2) $\mathcal{N}(\mathbf{a}) = \#(\mathbf{o}/\mathbf{a})$ $(\in \mathbf{N})$.

If $\mathbf{a} = (a)$ *is principal, then*

3) $\mathcal{N}(\mathbf{a}) = |N(a)|$ $(N(a) = a^{(1)} \cdot \ldots \cdot a^{(n)})$.

The Different. On our number field K we consider the **Q**-bilinear form

$$K \times K \longrightarrow \mathbf{Q} \quad , \quad (a, b) \longmapsto S(a \cdot b) \, .$$

It is non-degenerate, i.e.

$$S(ax) = 0 \quad \text{for all } x \in K \implies a = 0 \, .$$

Let $\mathbf{m} \subset K$ be a lattice. We define

$$\mathbf{m}^* = \{a \in K \mid S(ax) \in \mathbf{Z} \text{ for all } x \in \mathbf{m}\} \, .$$

AI.16. *If* $\mathbf{m} \subset K$ *is a lattice, then* $\mathbf{m}^*$ *is also a lattice of* K. *We have*

$$(\mathbf{m}^*)^* = \mathbf{m} \, .$$

If $\mathbf{m}$ *is an ideal, then* $\mathbf{m}^*$ *is also an ideal.*

We consider especially the dual lattice of the ring of integers. Its inverse is a distinguished integral ideal

$$\mathbf{d} = \mathbf{o}^{*-1} \subset \mathbf{o}$$

of K, the so-called **different** of K.

AI.17. *We have*

a) $\mathbf{a}^* \mathbf{a} = \mathbf{d}^{-1} = \mathbf{o}^*$,

b) $d(\mathbf{a}) = d_K \cdot \mathcal{N}(\mathbf{a})^2$,

c) $\mathcal{N}(\mathbf{d}) = |d_K|$.

The notion of the dual module is connected with the notion of the dual of a lattice $\mathbf{t} \subset \mathbf{R}^n$, which is defined as

$$\mathbf{t}^0 = \{a \in \mathbf{R}^n \mid \sum_{j=1}^n a_j x_j \in \mathbf{Z} \text{ for } x \in \mathbf{t}\}.$$

For sake of simplicity we explain this only for totally real number fields K. In this case we have n real imbeddings which give us

$$K \longrightarrow \mathbf{R}^n$$
$$a \longmapsto (a^{(1)}, \ldots, a^{(n)}).$$

AI.18. *The image* $\mathbf{n} \subset \mathbf{R}^n$ *of a lattice* $\mathbf{m} \subset K$ *is a lattice of* $\mathbf{R}^n$. *The image of the dual module* $\mathbf{m}^*$ *is the dual lattice* $\mathbf{n}^0$.

If we identify $\mathbf{m}$ with $\mathbf{n}$ we may write this simply as

$$\mathbf{m}^* = \mathbf{m}^0.$$

An Approximation Theorem. Let $\mathbf{t} \subset \mathbf{R}^n$ be a lattice. A **multiplier** of $\mathbf{t}$ is an element

$$\varepsilon \in \mathbf{R}_+^n \qquad (\mathbf{R}_+ = \{t \in \mathbf{R} \mid t > 0\})$$

with the property

$$\varepsilon \mathbf{t} = \mathbf{t}.$$

A multiplier always has norm 1

$$N\varepsilon := \varepsilon_1 \cdots \varepsilon_n = 1.$$

Let $\Lambda \subset \mathbf{R}_+^n$ be a discrete (multiplicative) subgroup of multipliers of maximal possible rank, i.e.

$$\Lambda \cong \mathbf{Z}^{n-1}.$$

This means that

$$\log \Lambda = \{(\log \varepsilon_1, \ldots, \log \varepsilon_n) \mid \varepsilon \in \Lambda\}$$

is a lattice in the hypersurface defined by

$$Sx = x_1 + \cdots + x_n = 0$$

in $\mathbf{R}^n$.

Example: Let K be a number field of degree n which is **totally real**, which means that each of the n imbeddings of K in $\mathbf{C}$ is real. We hence have an imbedding

$$K \hookrightarrow \mathbf{R}^n$$
$$a \mapsto (a^{(1)}, \ldots, a^{(n)}).$$

We denote by t the image of the ring of integers of K and by Λ the image of the group of all totally positive units in $\mathbf{R}^n$. By the Dirichlet unit theorem we actually have $\Lambda \cong \mathbf{Z}^{n-1}$. More generally one can take an arbitrary lattice $\mathbf{m} \subset K$ instead of $\mathbf{o}$ and a subgroup of finite index of Λ which acts on $\mathbf{m}$ by multiplication. Finally one can multiply the image of $\mathbf{m}$ by a real constant. It can be shown that each pair (t, Λ) arises in such a way.

AI.19 Proposition. *Let* $t \subset \mathbf{R}^n$ *be a lattice which admits a discrete group* $\Lambda \subset \mathbf{R}_+^n$ *of multipliers of maximal possible rank* $n - 1$. *We then have*

    1)   Each of the n projections

$$t \to \mathbf{R}, \ a \mapsto a_j \quad (1 \le j \le n),$$

is injective.

    2)   The image of t in $\mathbf{R}^{n-1}$ under each of the n projections

$$t \to \mathbf{R}^{n-1} \quad \textit{(cancellation of one component)}$$

is dense in $\mathbf{R}^{n-1}$.

Proof. 1) Assume that a is an element of t with the properties

$$a_1 = 0, \ a \ne 0.$$

We choose a multiplier $\varepsilon \in \Lambda$ with the property

$$\varepsilon_2 < 1, \ldots, \varepsilon_n < 1.$$

The existence of such a multiplier follows from the fact that

$$\{(\log \varepsilon_2, \ldots, \log \varepsilon_n) \mid \varepsilon \in \Lambda\}$$

is a lattice in $\mathbf{R}^{n-1}$. Now

$$a\varepsilon^n, \ n \in \mathbf{N},$$

is a sequence of non-zero elements of t which tends to 0. This contradicts the discreteness of t.

    2) The proof depends on the following trivial remark:

Remark. *Let* $t \subset \mathbf{R}^n$ *be a lattice. There exists a constant $r > 0$ such that each Euclidean ball of radius r contains a lattice point.*

For the proof of 2) we now consider an arbitrary open non-empty subset $U \subset \mathbf{R}^{n-1}$. We have to prove the existence of a lattice point in $\mathbf{R} \times U$. We choose a multiplier $\varepsilon \in \Lambda$ with

$$\varepsilon_2 > 1, \ldots, \varepsilon_n > 1.$$

It is sufficient to prove the existence of a lattice point in $\varepsilon^n \cdot (\mathbf{R} \times U)$ for some natural number n. But for sufficiently large n this set contains a Euclidean ball of arbitrarily large radius. The proof now follows from the remark. $\square$

A Zeta Function. We again consider a lattice $\mathbf{t} \subset \mathbf{R}^n$ together with a group $\Lambda \subset \mathbf{R}_+^n$ of multipliers of rank $n - 1$. A sign vector $\sigma = (\sigma_1, \ldots, \sigma_n)$ is a vector whose components are ± 1. There are precisely 2^n sign vectors.

Notation.

$$\mathbf{t}_\sigma = \{a \in \mathbf{t} \mid a\sigma > 0\} \, .$$

The group Λ acts on $\mathbf{t}_\sigma$.

AI.20 Proposition. *The series*

$$f_\sigma(s) = \sum_{a : \mathbf{t}_\sigma / \Lambda} |Na|^{-s}$$

converges if $s > 1$. The limit

$$\lim_{s \to 1+} (s - 1) f_\sigma(s)$$

exists and is independent of the sign vector σ.

We should mention that a deeper result of Hecke states that $(s - 1)f_\sigma(s)$ has an analytic continuation as holomorphic function into the whole s-plane.

A proof of AI.20 can be given by comparing the series with the integral

$$\int_B |Nx|^{-s} \, dx_1 \cdots dx_n \, ,$$

where B denotes a fundamental domain of Λ acting on

$$\{x \in \mathbf{R}^n \mid x\sigma > 0, \ |Nx| \geq 1\} \, .$$

We omit the details of this proof (which are not quite trivial). An immediate consequence of AI.20 is

AI.21 Corollary. *The limit*

$$\lim_{s \to 1+} \sum_{a : (\mathbf{t} - \{0\}) / \Lambda} \mathrm{sgn}\,(Na) |Na|^{-s}$$

exists.

This series occurs in Chap. II, §3 as "Shimizu's L-series".

II. Integration

We recall some basic facts about integration.

Let X be a locally compact topological space with a countable basis of its topology. We denote by

$$C_c(X) = \{f : X \to \mathbf{C} \text{ continuous} \mid \mathrm{supp}(f) \text{ compact}\},$$

where

$$\mathrm{supp}(f) = \text{closure of } \{x \in X \mid f(x) \neq 0\}$$

is the linear space of all compactly supported continuous functions on X.

A (Radon) measure on X is a $\mathbf{C}$-linear functional

$$I : C_c(X) \longrightarrow \mathbf{C}$$

such that

$$I(f) \geq 0 \text{ if } f \geq 0 \quad \text{(i.e. } f(x) \geq 0 \text{ for all } x \in X).$$

Notation.
$$I(f) = \int_X f(x)dx .$$

Basic example: $X = \mathbf{R}^n$, I the usual Riemann integral. The functional I can be extended to a larger class of functions, the so-called integrable functions (in the sense of Daniell-Lebesgue)

$$I : \mathcal{L}(X, dx) \longrightarrow \mathbf{C}$$
$$\mathcal{L}(X, dx) \supset C_c(X).$$

Before we formulate some of the basic properties of this integral we describe some derived notions.

A subset $A \subset X$ is called **neglectible** (with respect to dx), if the characteristic function

$$\chi_A(x) = \begin{cases} 1 & \text{for } x \in A \\ 0 & \text{for } x \notin A \end{cases}$$

is integrable (i.e. $\in \mathcal{L}(X, dx)$) and if its integral is 0.

AII.1. *Properties of neglectible sets:*

1) *Each subset of a neglectible set is neglectible.*
2) *The union of countably many neglectible sets is neglectible.*
3) *If f is an integrable function and g is an arbitrary function which differs from f only on a neglectible set, then also g is integrable and the integrals agree.*

4) *Assume $f \geq 0$. Then the two properties*

    a) $f \in \mathcal{L}(X, dx)$ and $I(f) = 0$

    b) $\{x \in X \mid f(x) \neq 0\}$ *is neglectible*

are equivalent.

A function $f : X \to \mathbb{C}$ is called **measurable** if for each pair of functions $\varphi, \psi \in C_c(X)$ the cutted function

$$f_{\varphi,\psi}(x) = \begin{cases} f(x) & \text{if } \varphi(x) \leq f(x) \leq \psi(x) \\ 0 & \text{elsewhere} \end{cases}$$

is integrable.

A subset $A \subset X$ is called measurable if its characteristic function is measurable. If it is integrable we call

$$\text{vol}(A) = \int \chi_A(x) dx$$

the **volume** of A (with respect to dx).

Notation. Let $f : A \to \mathbb{C}$ be a function on some subset A of X. Assume that the function

$$\tilde{f}(x) = \begin{cases} f(x) & \text{for } x \in A \\ 0 & \text{for } x \notin A \end{cases}$$

is integrable. We then say that f is **integrable along** A and write

$$\int_A f(x) dx := \int_X \tilde{f}(x) dx .$$

Special case:

$$\text{vol}(A) = \int_A dx .$$

AII.2. *Stability properties of measurable functions:*

1) *Continuous functions are measurable.*
2) *Integrable functions are measurable.*
3) *Open and closed subsets are measurable.*
4) *Sum and product of measurable functions are measurable.*
5) *If (f_n) is a sequence of measurable functions which converges pointwise to a function f, then f is also measurable.*
6) *The maximum and minimum of two measurable functions are measurable.*

Stability Properties of Integrable Functions.

AII.3. *The space* $\mathcal{L}(X, dx)$ *of integrable functions is a linear space over* **C**. *The integral is* **C**-*linear. Let* f *be any measurable function and* h *an integrable function with*

$$| f(x) | \leq h(x) \quad \text{for all } x \in X .$$

Then f *is also integrable and we have*

$$\left| \int_X f(x)dx \right| \leq \int_X h(x)dx .$$

Corollary.
$$f \in \mathcal{L}(X, dx) \implies | f | \in \mathcal{L}(X, dx) .$$

There are two fundamental theorems about the compatibility of the integral with limits.

AII.4. *Theorem of Beppo Levi:*
    Let
$$f_1(x) \leq f_2(x) \leq \cdots$$

be an increasing sequence of integrable functions. Assume that the sequence of their integrals is bounded:

$$\int_X f_n(x)dx \leq C .$$

Then there is an integrable function $f : X \to \mathbf{C}$ *such that*

$$f(x) = \lim_{n \to \infty} f_n(x)$$

outside some neglectible set. We have

$$\int_X f(x)dx = \lim_{n \to \infty} \int_X f_n(x)dx .$$

AII.5. *Theorem of Lebesgue:*
Let (f_n) *be a sequence of integrable functions which converges pointwise to a function* f. *Assume that there exists an integrable function* h *with*

$$| f_n(x) | \leq | h(x) | \quad \text{for all } x \in X \text{ and all } n \in \mathbf{N} .$$

Then f is integrable, and we have

$$\lim_{n\to\infty} \int_X f_n(x)dx = \int_X f(x)dx \; .$$

Product Measures. Let $(X, dx), (Y, dy)$ be two spaces with Radon measures. If $f : X \times Y \to \mathbf{C}$ is a continuous function with compact support on the product space, then one easily shows:

  a) The function

$$x \longmapsto f(x,y) \quad (y \text{ fixed})$$

is contained in $C_c(X)$.

  b) The function

$$y \longmapsto \int_X f(x,y)dx$$

is contained in $C_c(Y)$.

One can therefore define a Radon measure (the product measure) by the formula

$$\int_{X\times Y} f(x,y)dx\, dy = \int_Y \left[\int_X f(x,y)dx \right] dy \; .$$

AII.6. *Theorem of Fubini:*

For a function $f : X \times Y \to \mathbf{C}$ the following three statements are equivalent:

1) f is integrable (with respect to dx dy).
2) There is a neglectible set $S \subset Y$ such that

$$f_y : X \longrightarrow \mathbf{C}$$
$$f_y(x) = f(x,y)$$

is integrable for all $y \notin S$. The function

$$y \longmapsto \begin{cases} \int_X | f(x,y) | \, dx & \text{for } y \notin S \\ 0 & \text{for } y \in S \end{cases}$$

is integrable.
3) The same condition as in 2) with the roles of X and Y interchanged.

If these conditions are satisfied, then the formula

$$\int_{X\times Y} f(x,y)dx dy = \int_X \left[\int_Y f(x,y)dx \right] dy$$

$$= \int_Y \left[\int_X f(x,y)dy \right] dx$$

holds.

(The inner integrals have to be defined as 0 on some neglectible set.)

Quotient Measures. Let Γ be a subgroup of the group of all topological mappings of X onto itself. We assume that Γ acts discontinuously, i.e. for any compact subset $K \subset X$ the set of all

$$\gamma \in \Gamma \quad , \quad \gamma(K) \cap K \neq \emptyset,$$

is finite. Γ is especially countable because X may be written as a countable union of compact subsets. The quotient space X/Γ (equipped with the quotient topology) is again a locally compact space with countable topology.

Assumption. *The measure dx on X is Γ-invariant, i.e.*

$$\int_X f(x)dx \;=\; \int_X f(\gamma x)dx \quad for \; \gamma \in \Gamma.$$

Under this assumption we are going to construct a **quotient measure** $d\bar{x}$ on X/Γ.

Convention: There is a one-to-one correspondence between (continuous) functions on X/Γ and Γ-invariant (continuous) functions on X. We occasionally identify them.

There is a well-defined mapping

$$C_c(X) \longrightarrow C_c(X/\Gamma)$$
$$f(x) \longmapsto F(x) = \sum_{\gamma \in \Gamma} f(\gamma x).$$

The sum is locally finite.

It is not very hard to prove the following facts:

a) The mapping $f \mapsto F$ is surjective. If $F \geq 0$, one may achieve $f \geq 0$.
b) We have

$$F = 0 \implies \int_X f(x)dx = 0.$$

The properties a) and b) allow us to define a Radon measure $d\bar{x}$ on X/Γ by means of the formula

$$\int_{X/\Gamma} F(x)d\bar{x} \;=\; \int_X f(x)dx.$$

This formula extends to the class of integrable functions. (The situation is similar as in the theorem of Fubini.)

AII.7. *A subset $A \subset X/\Gamma$ is measurable (neglectible) if and only if its inverse image in X is measurable (neglectible). If $f : X \to \mathbf{C}$ is an integrable function (with respect to dx) the series*

$$F(x) = \sum_{\gamma \in \Gamma} f(\gamma x)$$

converges outside a certain Γ-invariant set S. If we define $F(x) = 0$ for $x \in S$ we have

$$F \in \mathcal{L}(X/\Gamma, d\bar{x})$$

and

$$\int_X f(x)dx = \int_{X/\Gamma} F(x)d\bar{x}$$

$$\left(= \int_{X/\Gamma} \sum_{\gamma \in \Gamma} f(\gamma x)d\bar{x} \right) .$$

Fundamental Domains. A measurable subset $F \subset X$ is called a **fundamental domain** with respect to Γ if

$$X = \bigcup_{\gamma \in \Gamma} \gamma(F)$$

and if there exists a neglectible set $S \subset F$ such that two different points in $F - S$ are never equivalent with respect to Γ.

AII.8. *A Γ-invariant function $f : X \to \mathbf{C}$ is integrable with respect to $d\bar{x}$ ($f \in \mathcal{L}(X/\Gamma, d\bar{x})$) if and only if it is integrable along F with respect to dx, and in this case we have*

$$\int_{X/\Gamma} f(x)d\bar{x} = \int_F f(x)dx .$$

The integral on the right-hand side is especially independent of the choice of a fundamental domain F.

Construction of Fundamental Domains. Let "$\sim$" be an equivalence relation on an arbitrary set X. If A is any subset of X, we denote by $\tilde{A}$ the set of all elements of X which are equivalent to an element of A.

Assume that a sequence

$$A_1, A_2, \ldots$$

of subsets of X with the following two properties is given:

a) $X = A_1 \cup A_2 \cup \ldots$

b) Two different elements of an A_j are never equivalent.

AII.9. *Under the above assumption the set*

$$F = F_1 \cup F_2 \cup \ldots$$

where

$$F_k = A_k - (A_1 \cup \overset{\frown}{\ldots} \cup A_{k-1})$$

is a set of representatives with respect to the given equivalence relation.

We apply this trivial remark and construct a measurable fundamental domain. In all cases considered in this book the set of fixed points

$$S = \{a \in X \mid \Gamma_a \neq \{\mathrm{id}\}\}$$

is a closed neglectible set. We may replace X by $X - S$ and therefore assume that Γ acts freely ($S = \emptyset$). We denote by $U_1, U_2, \ldots$ a countable basis of the topology (i.e each open set is a union of U_j's). We may assume that for all j

$$U_j \cap \gamma(U_j) = \emptyset \quad \text{for} \quad \gamma \in \Gamma - \{\mathrm{id}\}\,,$$

because Γ acts freely on X.

Let dx be a Radon measure on X with the following property. If A is a measurable set, then $\gamma(A)$ is also measurable for $\gamma \in \Gamma$. This is for example the case if dx is Γ-invariant. Under this condition the hull

$$\tilde{A} = \bigcup_{\gamma \in \Gamma} \gamma(A)$$

of a measurable set is measurable. The construction AII.9 with U_j instead of A_j gives us a measurable fundamental domain.

If $F_0 \subset X$ is a measurable fundamental set, i.e.

$$X = \bigcup_{\gamma \in \Gamma} \gamma(F_0)$$

(not necessarily disjoint), we may apply the above construction to $U_j \cap F_0$ instead of U_j, and we obtain

AII.10. *Under the above assumption each measurable fundamental set contains a fundamental domain.*

III. Alternating Differential Forms

We give a brief introduction into the theory of alternating differential forms.

We denote by

$$\mathcal{M}_p = \mathcal{M}_p^{(n)} = \{a \subset \{1,\ldots,n\} \mid \#a = p\}$$

the set of all subsets a of $\{1,\ldots,n\}$ which consist of p elements. Their number is

$$\binom{n}{p} \qquad (:= 0 \text{ if } p < 0 \text{ or } p > n).$$

Definition. *A (n alternating) differential form ω of degree p (shortly p-form) on an open domain $D \subset \mathbb{R}^n$ is a system of functions*

$$(f_a)_{a \in \mathcal{M}_p} \quad , \quad f_a : D \longrightarrow \mathbb{C}.$$

Notation.

$M^p(D) :=$ set of all differential forms of degree p.

$M_\infty^p(D) :=$ set of all C^∞-differential forms of degree p (i.e. all components f_a are C^∞-functions) $(= 0$ if $p > n$ or $p < 0)$.

We may identify $M^0(D)$ with the set of functions on D (because $\mathcal{M}_0$ consists of one element, namely the empty set). We have especially

$$M_\infty^0(D) = C^\infty(D).$$

We give the basic definitions of the calculus of differential forms:

I. $M^p(D)$ is an $M^0(D)$-module:

$$(f_a) + (g_a) = (f_a + g_a),$$
$$f \cdot (f_a) = (f \cdot f_a).$$

II. The Total Differential of a Function. We may identify

$$M^1(D) = M^0(D)^n = M^0(D) \times \ldots \times M^0(D)$$
$$M_\infty^1(D) = C^\infty(D)^n.$$

We define the total differential by

$$d : C^\infty(D) \longrightarrow M^1(D)$$
$$df = (\frac{\partial f}{\partial x_1}, \ldots, \frac{\partial f}{\partial x_n}).$$

One has

a)
$$d(f + g) = df + dg$$

b)
$$d(fg) = fdg + gdf$$

c)
$$df = 0 \iff f \text{ is locally constant}.$$

Notation.
$$dx_j := (0, \ldots, 0, 1, 0, \ldots, 0)$$

(This is the differential of the function $p_j(x) = x_j$).

With this notation we may write

$$df = \sum_{j=1}^{n} \frac{\partial f}{\partial x_j} dx_j.$$

III. The Alternating Product. In analogy to the case $p = 1$ we define the p-form dx_a for a subset $a \subset \{1, \ldots, n\}$ of order p by

$$(dx_a)_b = \begin{cases} 1 & \text{if } a = b, \\ 0 & \text{if } a \neq b. \end{cases}$$

We may write an arbitrary p-form $\omega = (f_a)$ as

$$\omega = \sum_{a \in \mathcal{M}_p} f_a dx_a.$$

Let $a, b \subset \{1, \ldots, n\}$ be two subsets of order p. We define a certain "sign"

$$\varepsilon(a, b) \in \{0, 1, -1\}.$$

Case 1:
$$\varepsilon(a, b) = 0 \quad \text{if} \quad a \cap b \neq \emptyset,$$

Case 2: Assume $a \cap b = \emptyset$.

Let $a_1, \ldots, a_p$ be the elements of a in their natural order:

$$a: a_1 < \ldots < a_p$$

and analogous

$$b: b_1 < \ldots < b_q.$$

We have

$$a \cup b = \{a_1, \ldots, a_p, b_1, \ldots, b_q\}.$$

The elements in the brackets are not necessarily in their natural order. We denote by $\varepsilon(a, b)$ the sign of the permutation which arranges them in their natural order. We now define

$$dx_a \wedge dx_b = \varepsilon(a, b) dx_{a \cup b}$$

and extend this bilinearly to a mapping

$$M^p(D) \times M^q(D) \longrightarrow M^{p+q}(D),$$

$$\left(\sum_a f_a dx_a\right) \wedge \left(\sum_b g_b dx_b\right) = \sum_{a,b} f_a g_b \, dx_a \wedge dx_b.$$

It is easy to verify the following rules

1)
$$f \wedge \omega = f \cdot \omega \quad \text{if} \quad f \in M^0(D),$$

2)
$$\omega \wedge \omega' = (-1)^{pq} \omega' \wedge \omega \quad \text{if} \quad \omega \in M^p(D), \, \omega' \in M^q(D),$$

especially
$$\omega \wedge \omega = 0 \quad \text{if} \quad p \text{ is odd},$$

3)
$$(\omega \wedge \omega') \wedge \omega'' = \omega \wedge (\omega' \wedge \omega''),$$

4) $(\omega_1 + \omega_2) \wedge \omega = \omega_1 \wedge \omega + \omega_2 \wedge \omega \quad \text{if} \quad \omega_1, \omega_2 \in M^p(D), \, \omega \in M^q(D).$

Because of the law of associativity 3) we may define the alternating product $\omega_1 \wedge \ldots \wedge \omega_p$ of several alternating differential forms. We obviously have

$$dx_a = dx_{a_1} \wedge \ldots \wedge dx_{a_p}$$
$$a = \{a_1, \ldots, a_p\} \quad , \quad a_1 < \ldots < a_p.$$

We may therefore write a differential form in the following form

$$\omega = \sum_{a \in M_p} f_a dx_a$$
$$= \sum_{1 \leq a_1 < \ldots < a_p \leq n} f_{a_1, \ldots, a_p} dx_{a_1} \wedge \ldots \wedge dx_{a_p}.$$

Example: The alternating product of two 1-forms (= differential forms of degree 1) is given by

$$\left(\sum_{\nu=1}^n f_\nu dx_\nu\right)\left(\sum_{\mu=1}^n g_\mu dx_\mu\right) = \sum_{1 \leq \nu < \mu \leq n} (f_\nu g_\mu - f_\mu g_\nu) dx_\nu \wedge dx_\mu,$$

because

$$dx_\nu \wedge dx_\mu = -dx_\mu \wedge dx_\nu \quad (= 0 \text{ if } \nu = \mu).$$

IV. Exterior Differential. We define

$$d : M_\infty^p(D) \longrightarrow M_\infty^{p+1}(D)$$

by the formula
$$d(\sum f_a dx_a) = \sum df_a \wedge dx_a.$$

It is easy to verify the following rules:

1) $$d(\omega + \omega') = d\omega + d\omega',$$

2) $d(\omega \wedge \omega') = (d\omega) \wedge \omega' + (-1)^p \omega \wedge d\omega' \quad, \quad \omega \in M_\infty^p(D),\ \omega' \in M_\infty^q(D),$

3) $$d(d\omega) = 0.$$

As a special case of 2) we get the formula
$$d(\omega \wedge \omega') = d\omega \wedge \omega' \quad \text{if} \quad d\omega' = 0.$$

We obtain by induction
$$d(\omega \wedge df_1 \wedge \ldots \wedge df_m) = d\omega \wedge df_1 \wedge \ldots \wedge df_m.$$

V. Transformation of Differential Forms. Let
$$\varphi : D \longrightarrow D'$$

be a C^∞-mapping between open domains $D \subset \mathbf{R}^n$, $D' \subset \mathbf{R}^m$. We denote the co-ordinates of D by $x = (x_1, \ldots, x_n)$ and those of D' by $y = (y_1, \ldots, y_m)$. We want to construct a mapping
$$M^p(D') \longrightarrow M^p(D)$$
$$\omega \longmapsto \varphi^*\omega$$

such that the following "axioms" hold:

1) In the case of functions $(p = 0)$ we have
$$\varphi^*(f) = f \circ \varphi.$$

2) $$\varphi^*(\omega + \omega') = \varphi^*\omega + \varphi^*\omega'.$$

3) $$\varphi^*(\omega \wedge \omega') = \varphi^*\omega \wedge \varphi^*\omega'.$$

4) $$\varphi^*(d\omega) = d(\varphi^*\omega) \quad (\omega \in M_\infty^p(D)).$$

Of course such a mapping is unique: From 4) we obtain
$$\varphi^*(dy_i) = d\varphi_i = \sum_{j=1}^n \frac{\partial \varphi_i}{\partial x_j} dx_j$$

or

$$\varphi^*(\sum f_i dy_i) = \sum g_j dx_j \,,$$

where

$$\begin{pmatrix} g_1(x) \\ \vdots \\ g_n(x) \end{pmatrix} = \mathcal{J}(\varphi, x)' \begin{pmatrix} f_1(\varphi(x)) \\ \vdots \\ f_m(\varphi(x)) \end{pmatrix} \,.$$

Here $\mathcal{J}(\varphi, x)'$ is the transpose of the Jacobian

$$\mathcal{J}(\varphi, x) = \begin{pmatrix} \partial\varphi_1/\partial x_1 & \cdots & \partial\varphi_1/\partial x_n \\ \vdots & & \vdots \\ \partial\varphi_m/\partial x_1 & \cdots & \partial\varphi_m/\partial x_n \end{pmatrix} \,.$$

A very important case is

$$p = m = n \,.$$

A differential form of "top degree" $p = n$ on D' is of the form

$$\omega = f dy_1 \wedge \ldots \wedge dy_n \,.$$

We have

$$\varphi^*\omega = g dx_1 \wedge \ldots \wedge dx_n$$

with a certain function g. From the characterization of the determinant as an alternating multilinear form of the rows with certain properties one obtains

$$g(x) = \det \mathcal{J}(\varphi, x) f(\varphi(x)) \,.$$

VI. Complex Co-ordinates. We now consider the case of an open domain $D \subset \mathbf{C}^n$. We may identify $\mathbf{C}^n$ with $\mathbf{R}^{2n}$ by means of

$$(z_1, \ldots, z_n) \longleftrightarrow (x_1, y_1, \ldots, x_n, y_n)$$

and therefore apply the "real theory" of differential forms to this case. But it is frequently useful to use complex co-ordinates instead of real ones, for example

$$dz_\nu := dx_\nu + i\, dy_\nu \,,$$
$$d\bar{z}_\nu := dx_\nu - i\, dy_\nu$$

instead of

$$dx_j = (dz_j + d\bar{z}_j)/2 \,,$$
$$dy_j = (dz_j - d\bar{z}_j)/2i \,.$$

Notation.

$$dz_a := dz_{a_1} \wedge \ldots \wedge dz_{a_p} \, ,$$

$$d\bar{z}_a := d\bar{z}_{a_1} \wedge \ldots \wedge d\bar{z}_{a_p} \, ,$$

$$a = \{a_1, \ldots, a_p\} \, , \quad 1 \le a_1 < \ldots < a_p \le n \, ,$$

$$M^{p,q}(D) := \sum_{\#a=p, \#b=q} M^0(D) dz_a \wedge d\bar{z}_b$$

with $a, b \subset \{1, \ldots, n\}$.

We obviously have a decomposition

$$M^m(D) = \sum_{p+q=m} M^{p,q}(D)$$

as a direct sum, i.e. the decomposition of an $\omega \in M^m(D)$ into a sum

$$\omega = \sum_{p+q=m} \omega_{p,q} \, , \quad \omega_{p,q} \in M^{p,q}(D) \, ,$$

is unique (as well as the decomposition of a single $\omega_{p,q}$ into its components,

$$\omega_{p,q} = \sum_{\#a=p, \#b=q} f_{a,b} dz_a \wedge d\bar{z}_b \, .)$$

The alternating product respects this decomposition:

$$\omega \in M^{p,q}(D) \, , \ \omega' \in M^{p',q'}(D) \implies \omega \wedge \omega' \in M^{p+p',q+q'}(D) \, .$$

The exterior differentiation d does not respect the decomposition. To improve this situation, we write d as a sum of two operators $\partial, \bar{\partial}$ which do respect the decomposition: We define

$$\partial/\partial z_j := (\partial/\partial x_j - i\partial/\partial y_j)/2 \, ,$$
$$\partial/\partial \bar{z}_j := (\partial/\partial x_j + i\partial/\partial y_j)/2$$

and

$$\partial : C^\infty(D) \longrightarrow M^{1,0}_\infty(D) \, ,$$
$$\bar{\partial} : C^\infty(D) \longrightarrow M^{0,1}_\infty(D)$$

by

$$\partial f := \sum_{j=1}^{n}(\partial f/\partial z_j)dz_j\,,$$

$$\overline{\partial} f := \sum_{j=1}^{n}(\partial f/\partial \overline{z}_j)d\overline{z}_j\,.$$

More generally we define linear mappings

$$\partial : M_{\infty}^{p,q}(D) \longrightarrow M_{\infty}^{p+1,q}(D)$$
$$\overline{\partial} : M_{\infty}^{p,q}(D) \longrightarrow M_{\infty}^{p,q+1}(D)$$

by

$$\partial(fdz_a \wedge d\overline{z}_b) := (\partial f) \wedge dz_a \wedge d\overline{z}_b$$
$$\overline{\partial}(fdz_a \wedge d\overline{z}_b) := (\overline{\partial} f) \wedge dz_a \wedge d\overline{z}_b\,.$$

One can verify the rules

1) $$d = \partial + \overline{\partial}\,,$$

2) $$\partial(\omega \wedge \omega') = (\partial\omega) \wedge \omega' + (-1)^p \omega \wedge \partial\omega'\,,$$

3) $$\partial \circ \partial = 0\,, \quad \overline{\partial} \circ \overline{\partial} = 0\,, \quad \partial \circ \overline{\partial} = -\overline{\partial} \circ \partial\,.$$

VII. Holomorphic Differential Forms. A differential form ω on an open domain $D \subset \mathbf{C}^n$ is called **holomorphic** if the following two conditions hold

a) ω is of type $(p,0)$, i.e.

$$\omega = \sum f_{i_1,\dots,i_p}\, dz_{i_1} \wedge \dots \wedge dz_{i_p}\,,$$

b) The components $f_{i_1,\dots,i_p}$ are holomorphic.

Notice: A C^{∞}-function f is holomorphic iff $\overline{\partial} f = 0$ (Cauchy-Riemann differential equations). Therefore a C^{∞}-$(p,0)$-form ω is holomorphic iff

$$\overline{\partial}\omega = 0\,.$$

Notation.

$$\Omega^p(D) = \{\omega \in M^{p,0}(D) \mid \omega \text{ holomorphic}\}\,.$$

We have

$$d = \partial : \Omega^p(D) \longrightarrow \Omega^{p+1}(D)$$
$$(\partial(\sum f_a dz_a) = \sum \partial f_a \wedge dz_a\,)\,.$$

VIII. Holomorphic Transformation of Differential Forms. Let

$$\varphi : D \longrightarrow D'$$

be a holomorphic mapping between open domains $D \subset \mathbf{C}^n$, $D' \subset \mathbf{C}^m$. From the Cauchy-Riemann differential equations it follows that

$$\varphi^*(dw_i) = \sum (\partial \varphi_i / \partial z_j) dz_j$$
$$\varphi^*(d\overline{w}_i) = \sum (\overline{\partial \varphi_i / \partial z_j}) d\overline{z}_j \,.$$

We therefore obtain mappings

$$\varphi^* : M^{p,q}(D') \longrightarrow M^{p,q}(D) \quad (\varphi \text{ holomorphic!}) \,,$$
$$\varphi^* : \Omega^p(D') \longrightarrow \Omega^p(D) \,.$$

In the special case $n = m$ we have

$$\varphi^*(dw_1 \wedge \ldots dw_n) = j(\varphi, z) dz_1 \wedge \ldots \wedge dz_n$$

where $j(\varphi, z)$ is the determinant of the (complex) Jacobian,

$$j(\varphi, z) = \det(\partial \varphi_i / \partial z_j) \,.$$

IX. Riemannian Metric. A Riemannian metric on an open domain $D \subset \mathbf{R}^n$ is a symmetric $n \times n$-matrix

$$g = (g_{ik}) \quad, \quad g_{ik} : D \to \mathbf{R} \quad (1 \le i, k \le n) \,,$$

of real C^∞-functions on D such that $g(x)$ is positive definite for all $x \in D$.

Let

$$\varphi : D' \longrightarrow D$$

be a C^∞-mapping between open domains $D' \subset \mathbf{R}^m$ and $D \subset \mathbf{R}^n$. We define $\varphi^*(g)$ by

$$\varphi^*(g)(y) := \mathcal{J}'(\varphi, y) g(\varphi(y)) \mathcal{J}(\varphi, y)$$
$$(\mathcal{J}(\varphi, x) \text{ Jacobian}) \,.$$

This is a symmetric real positive semidefinite matrix of C^∞-functions. It defines a Riemannian metric, if φ is a local immersion, i.e. $\mathcal{J}(\varphi, y)$ has rank m for all $y \in D$.

A diffeomorphism

$$\varphi : D \xrightarrow{\sim} D$$

is called a **motion** with respect to g if

$$\varphi^* g = g \,.$$

Our basic example is the upper half plane equipped with the so-called Poincaré metric

$$g(z) = \begin{pmatrix} y^{-2} & 0 \\ 0 & y^{-2} \end{pmatrix} \,.$$

It is easy to prove that the fractional linear transformations

$$z \longmapsto Mz, \quad M \in SL(2,\mathbf{R}),$$

are motions with respect to g. We want to use the metric g on $D \subset \mathbf{R}^n$ to construct a certain pairing

$$M^p(D) \times M^p(D) \longrightarrow M^0(D)$$
$$(\omega, \omega') \longmapsto < \omega, \omega' > .$$

Here "pairing" means a $M^0(D)$-bilinear mapping. There exists a unique pairing

$$M^p(D) \times M^p(D) \longrightarrow M^0(D)$$

with the following properties:

a)
$$< f, g > = f \cdot g \quad \text{if} \quad f, g \in M^0(D).$$

b)
$$< dx_i, dx_k > = g^{ik},$$

where g^{ik} denotes the components of the inverse of g,

$$g^{-1} = (g^{ik}).$$

c)
$$< \omega_1 \wedge \ldots \wedge \omega_p, \omega_1' \wedge \ldots \wedge \omega_p' > = \det(< \omega_i, \omega_j' >)_{1 \le i, j \le p} \quad (\omega_i, \omega_i' \in M^1(D)).$$

The proof is easy and straightforward. One defines the pairing on the elements of the basis $dx_{i_1} \wedge \ldots \wedge dx_{i_p}$, extends it bilinearly and proves the properties b) and c).

If $\varphi : D' \to D$ is a diffeomorphism we have

$$< \varphi^* \omega, \varphi^* \omega' >_{\varphi^* g} = \varphi^* < \omega, \omega' >_g .$$

We have especially: *The constructed pairing is invariant with respect to motions.*

The so-called **fundamental form** on an open domain $D \subset \mathbf{R}^n$ with respect to a Riemannian metric g is defined as

$$\omega_g := +\sqrt{\det g} \, dx_1 \wedge \ldots \wedge dx_n .$$

This fundamental form is also invariant with respect to a motion φ,

$$\varphi^* \omega_g = \omega_g ,$$

if φ **preserves the orientation**, i.e.

$$\det \mathcal{J}(\varphi, x) > 0 \quad \text{for all } x \in D .$$

(More generally we have for an orientation preserving diffeomorphism

$$\varphi^* : D' \to D$$
$$\varphi^* \omega_g = \omega_{\varphi^* g} .)$$

X. The Star Operator. Let $D \subset \mathbf{R}^n$ be an open domain with Riemannian metric g. There exists a unique $M^0(D)$-linear isomorphism

$$* : M^p(D) \xrightarrow{\;\sim\;} M^{n-p}(D)$$

such that

$$< *\omega, \omega' > \omega_g = \omega \wedge \omega'$$

for all

$$\omega \in M^p(D) \quad , \quad \omega' \in M^{n-p}(D) .$$

The star operator is *invariant with respect to orientation preserving motions* φ:

$$*(\varphi^* \omega) = \varphi^* (*\omega) .$$

One has

$$*(*\omega) = (-1)^{pn+p} \omega .$$

The **codifferentiation**

$$\delta : M^p_\infty(D) \longrightarrow M^{p-1}_\infty(D) \quad (= 0 \text{ if } p = 0)$$

is defined by

$$\delta := (-1)^{np+n+1} * d * .$$

The **Laplace-Beltrami operator** is

$$\Delta : M^p_\infty(D) \longrightarrow M^p_\infty(D) ,$$

$$\Delta := d\delta + \delta d .$$

Of course δ and Δ commute with orientation preserving motions. (More generally: If $\varphi : D' \to D$ is an orientation preserving diffeomorphism we have

$$\varphi^* \circ \Delta_g = \Delta_{\varphi^* g} .)$$

XI. Hermitean Metric. A Hermitean metric is a complex matrix h with the property $\overline{h}' = h$. Each $n \times n$-Hermitean matrix defines a real $2n \times 2n$ symmetric matrix g which is characterized by

$$\overline{z}' h z = a' g a ,$$

where

$$a' = (x_1, y_1, \ldots, x_n, y_n) .$$

We say that g comes from the Hermitean matrix h. A Hermitean metric on an open domain $D \subset \mathbf{C}^n$ is an $n \times n$-matrix h of C^∞-functions on D, such that $h(z)$ is Hermitean and positive definite for all $z \in D$. The associate $2n \times 2n$-matrix g is a Riemannian metric. We have

$$\det h(z) = +\sqrt{\det g(z)}\,.$$

The fundamental form may be written as

$$\omega_g = \sqrt{\det g}\; dx_1 \wedge dy_1 \wedge \ldots \wedge dx_n \wedge dy_n$$

$$= \frac{1}{(2i)^n}(\det h)dz_1 \wedge d\bar{z}_1 \wedge \ldots \wedge dz_n \wedge d\bar{z}_n\,.$$

Let now

$$\varphi : D \overset{\sim}{\longrightarrow} D'\,, \quad D, D' \subset \mathbf{C}^n \text{ open}\,,$$

be a **biholomorphic mapping**, h a Hermitean metric on D' and g the associate Riemannian metric. Then the pulled back Riemannian metric $\varphi^* g$ on D comes from a Hermitean metric, namely from

$$(\varphi^* h)(z) := \overline{\mathcal{J}_{\mathbf{z}}(\varphi, z)}'\, h(\varphi(z))\mathcal{J}_{\mathbf{z}}(z, \varphi)\,.$$

Here $\mathcal{J}_{\mathbf{z}}(\varphi, z)$ denotes the **complex** Jacobian.

The Laplace-Beltrami operator

$$-\Delta = d * d * + * d * d$$

(the real dimension of $\mathbf{C}^n$ is even) usually does not preserve the decomposition

$$M^m(D) = \sum_{p+q=m} M^{p,q}(D)\,.$$

Therefore one also considers the operators

$$-\square = \partial * \bar{\partial} * + * \bar{\partial} * \partial$$
$$-\square = \bar{\partial} * \partial * + * \partial * \bar{\partial}$$

which map $M^{p,q}(D)$ into itself. This follows immediately from the following fact: In the case of a Hermitean metric the star operator maps (p,q)-forms to $(n-q, n-p)$-forms

$$* : M^{p,q}(D) \longrightarrow M^{n-q,n-p}(D)\,.$$

XII. Kählerian Metric. A Hermitean metric h on an open set $D \subset \mathbf{C}^n$ is called locally Euclidean at a point $a \in D$ if $h(a)$ is the unit matrix and if the first partial derivatives of h vanish at a.

Definition. *A Hermitean metric h is called Kählerian if it is locally equivalent with a Euclidean one, i.e.: For each point $a \in D$ there exists a biholomorphic mapping*

$$\varphi : U \overset{\sim}{\longrightarrow} V$$

of an open subset $U \subset \mathbf{C}^n$ onto an open neighbourhood V of a in D such that the pulled back metric $\varphi^* h$ is locally Euclidean at $b = \varphi^{-1}(a)$.

For any Hermitean metric h we may consider the (1,1)-form

$$\Omega = \Omega(h) = \frac{1}{2\pi i} \sum_{1 \le i,k \le n} h_{ik} dz_i \wedge d\bar{z}_k .$$

This differential form is invariant with respect to biholomorphic transformations φ, i.e.

$$\Omega(\varphi^* h) = \varphi^* \Omega(h) .$$

Remark. In the case of a Kählerian metric h we have

$$d\Omega = 0 .$$

The converse is also true but we do not need this.

Proposition. *In the case of a Kählerian metric we have the identities*

$$\Delta = 2\square = 2\,\overline{\square} \ .$$

Corollary. *The Laplace-Beltrami operator Δ preserves the double graduation*

$$M_\infty^m(D) = \sum_{p+q=m} M_\infty^{p,q}(D) .$$

We make use of this proposition in Chap. III, §1 and therefore we indicate the proof: Let us consider the operator

$$L : M^{p,q} \longrightarrow M^{p+1,q+1}$$
$$L(\omega) = \Omega \wedge \omega ,$$

where Ω denotes the Kähler form (see above). The identities in the proposition are a formal consequence of the relations

$$L \circ * \partial * \ - \ * \partial * \circ L \ = \ i\partial$$
$$L \circ * \bar{\partial} * \ - \ * \bar{\partial} * \circ L \ = \ -i\bar{\partial} .$$

We leave this reduction to the reader. The advantage of the latter relations is that they involve only first order derivatives. From the definition of the

Kähler property it follows that such a relation, which is invariant under biholomorphic transformations, has to be proved only in the case of the Euclidean metric $h = E =$ unit matrix. In this case the relations can be verified by direct calculation.

Example. *Each Hermitean metric h on a (complex) 1-dimensional domain $D \subset \mathbb{C}$ is Kählerian.*

The de Rham Complex

XIII. Differential Forms on Manifolds. Let X be a topological space. We always assume that X is a Hausdorff space with countable basis of its topology. A differentiable structure on X is a family

$$\varphi_j : U_j \longrightarrow V_j, \quad U_j \subset X \text{ open}, \ V_j \subset \mathbb{R}^n \text{ open},$$

of topological mappings with the properties

a) $$X = \bigcup U_i,$$

b) $$\varphi_j \circ \varphi_i^{-1} : \varphi_i(U_i \cap U_j) \longrightarrow \varphi_j(U_i \cap U_j)$$

is a C^∞-diffeomorphism for all (i,j).

The space X together with a distinguished differentiable structure is called a differentiable manifold of dimension n.

A differential form of degree p on X is a family

$$\omega = (\omega_i), \quad \omega_i \in M^p(V_i),$$

such that the formula

$$(\varphi_j \circ \varphi_i^{-1})^* \omega_j = \omega_i$$

holds on $\varphi_i(U_i \cap U_j)$. We denote by $M^p(X)$ the space of all p-forms on X and by $M_\infty^p(X)$ the space of all C^∞-p-forms (i.e. all ω_i are C^∞). We may identify a function $f : X \to \mathbb{C}$ with the zero form

$$(f_i), \quad f_i = f \,|\, U_i \circ \varphi_i^{-1}.$$

The function f is called C^∞-differentiable if all the f_i's are C^∞-differentiable. We hence may identify $M_\infty^0(X)$ and

$$C^\infty(X) = \{f : X \to \mathbb{C} \mid f \text{ is } C^\infty\}.$$

There are natural mappings

$$M^p(X) \times M^q(X) \longrightarrow M^{p+q}(X)$$

$$(\omega, \omega') \longmapsto \omega \wedge \omega',$$

$$(\omega \wedge \omega')_i := \omega_i \wedge \omega'_i$$

and

$$M^p_\infty(X) \xrightarrow{\ d\ } M^{p+1}_\infty(X),$$
$$(d\omega)_i := d\omega_i .$$

The sequence

$$\ldots \longrightarrow M^p_\infty(X) \xrightarrow{\ d\ } M^{p+1}_\infty(X) \longrightarrow \ldots$$

is the so-called **de Rham complex** (complex means: $d \circ d = 0$).

The de Rham cohomology groups (they are actually vector spaces over
C) are defined as

$$H^p(X) := C^p(X)/B^p(X)$$

where

$$C^p(X) = \ker(M^p_\infty(X) \xrightarrow{\ d\ } M^{p+1}_\infty(X))$$

$$B^p(X) = \operatorname{im}(M^{p-1}_\infty(X) \xrightarrow{\ d\ } M^p_\infty(X)) .$$

By the theorem of de Rham there exists a natural isomorphism

$$H^p(X) \xrightarrow{\ \sim\ } H^p(X, \mathbf{C})$$

where $H^p(X, \mathbf{C})$ denotes the singular cohomology groups with coefficients in **C**.

XIV. Real Hodge Theory. The Hodge theory is a powerful tool to compute
the de Rham cohomology groups in the case of a compact manifold:

A **Riemannian metric** g on the differentiable manifold X is a family

$$g = (g_i), \quad g_i \text{ Riemannian metric on } V_i$$

such that the transformation formula

$$(\varphi_j \circ \varphi_i^{-1})^* g_j = g_i$$

is valid on $\varphi_i(U_i \cap U_j)$. If a Riemannian metric is given, the star operator

$$* : M^p(X) \longrightarrow M^{n-p}(X)$$
$$(*\omega)_i := *(\omega_i)$$

is well defined. We therefore may define the Laplace-Beltrami operator

$$\Delta : M_\infty^p(X) \longrightarrow M_\infty^p(X)$$
$$(\Delta\omega)_i = \Delta\omega_i .$$

The kernel of Δ is the space of harmonic forms.

$$\mathcal{H}^p(X) = \ker\left(M_\infty^p(X) \xrightarrow{\Delta} M_\infty^p(X)\right) .$$

One of the main results of the real Hodge theory states:

Assume that X is compact. Then each harmonic form is closed. The natural mapping

$$\mathcal{H}^p(X) \longrightarrow H^p(X)$$

is an isomorphism.

Notice: If ω is harmonic then $*\omega$ is also harmonic. We obtain (for a compact manifold!)

$$\omega \text{ harmonic} \iff d\omega = 0 \text{ and } d(*\omega) = 0 .$$

XV. Integration of n-forms.

An n-form

$$\omega = f dx_1 \wedge \ldots \wedge dx_n$$

on an open domain $D \subset \mathbf{R}^n$ is called integrable if the function f is integrable with respect to the Euclidean measure:

Notation.

$$\int_D \omega := \int_D f(x) dx_1 \ldots dx_n .$$

If

$$\varphi : D' \longrightarrow D$$

is an orientation preserving diffeomorphism, we have

$$\int_{D'} \varphi^*\omega = \int_D \omega .$$

We hence may generalize the notion of an integrable n-form ω and the value

$$\int_X \omega$$

to an arbitrary **oriented** differentiable manifold. Here "oriented" means that all transition functions $\varphi_j \circ \varphi_i^{-1}$ are orientation preserving.

A differential form ω of arbitrary degree p on an oriented Riemannian manifold (X, g) is called **square integrable**, if the n-form ($n = \dim X$)

$$\omega \wedge *\overline{\omega}$$

is integrable.

XVI. Some Results on Non-compact Manifolds.

Theorem. *Let ω be a square integrable and closed $(d\omega = 0)$ C^∞-differential form on an oriented Riemannian manifold. There exists a square integrable harmonic form ω_0 such that*

$$\omega = \omega_0 + d\tilde{\omega}.$$

($\tilde{\omega}$ some C^∞-differential form.)

But in contrast to the compact case the form ω_0 needs not to be unique and not each square integrable harmonic form needs to be closed.

But there is a very remarkable

Theorem. *Let (X, g) be an oriented complete Riemannian manifold. Then each square integrable harmonic form is closed.*

What does complete mean?

Let

$$\alpha : [0, 1] \longrightarrow D$$

be a C^∞-differentiable curve in a domain $D \subset \mathbf{R}^n$ which is equipped with a Riemannian metric. The **velocity** of α at $t \in [0, 1]$ is defined by

$$v_\alpha(t) = \sum_{1 \le i,j \le n} g_{ij}\big(\alpha(t)\big)\dot{\alpha}_i(t)\dot{\alpha}_j(t).$$

The length of α is defined by

$$\ell(\alpha) = \ell_g(\alpha) = \int_0^1 \sqrt{v_\alpha(t)}\,dt.$$

The notion of velocity and arc length is invariant with respect to diffeomorphisms $\varphi : D' \to D$ if one replaces α by $\alpha \circ \varphi$ and g by $\varphi^* g$. One may especially generalize these notions to the case of curves

$$\alpha : [0, 1] \longrightarrow X$$

in an arbitrary Riemannian manifold (X, g). One defines a metric on a connected (X, g) by means of

$$d(a, b) := \inf_\alpha \ell(\alpha),$$

where α runs over all curves connecting a and b.

(X, g) is called **complete**, if each Cauchy sequence with respect to this metric converges.

XVII. Complex Hodge Theory. A complex analytic manifold of (complex) dimension n is a topological space X together with a family

$$\varphi_i : U_i \longrightarrow V_i \quad U_i \subset X \text{ open}, \quad V_i \subset \mathbf{C}^n \text{ open}$$

of topological mappings such that

a) $X = \bigcup U_i$

b) all the transition maps $\varphi_j \circ \varphi_i^{-1}$ are biholomorphic.

Each complex analytic manifold is also an oriented differentiable manifold of (real) dimension $2n$. We may also consider the Dolbeault complex

$$\cdots \longrightarrow M^{p,q}(X) \overset{\bar\partial}{\longrightarrow} M^{p,q+1}(X) \longrightarrow \cdots$$

and the Dolbeault cohomology groups

$$H^{p,q}(X) = C^{p,q}(X)/B^{p,q}(X),$$

where

$$C^{p,q}(X) = \ker(M^{p,q}(X) \overset{\bar\partial}{\longrightarrow} M^{p,q+1}(X))$$
$$B^{p,q}(X) = \operatorname{im}(M^{p,q-1}(X) \overset{\bar\partial}{\longrightarrow} M^{p,q}(X)).$$

Remark. There is a natural isomorphism

$$H^{p,q}(X) \cong H^q(X, \Omega^p_X),$$

where Ω^p_X denotes the sheaf of holomorphic p-forms on X.

Similar to the real case we may generalize the local theory of differential forms on complex domains to the case of analytic manifolds. We especially may define the notion of a (p,q)-form and have the decomposition

$$M^m(X) = \bigoplus_{p+q=m} M^{p,q}(X).$$

The same is true with the subscript "∞".

We now assume that a Hermitean metric h (i.e. a compatible family (h_i) of Hermitean metrics h_i on V_i) is given. Then the operators $\square$, $\bar\square$ are well defined. If h is Kählerian (i.e. all h_i are Kählerian) we have

$$\Delta = 2\square.$$

Notation.

$$\mathcal{H}^{p,q}(X) = \{\omega \in M^{p,q}_\infty(X) \mid \square\omega = 0\}.$$

In the Kählerian case we have

$$\mathcal{H}^m(X) = \bigoplus_{p+q=m} \mathcal{H}^{p,q}(X).$$

The main result of the complex Hodge theory is

Theorem. *Let X be a compact Kählerian manifold. Each harmonic form ω is $\bar{\partial}$-closed (i.e. $\bar{\partial}\omega = 0$). The natural mapping*

$$\mathcal{H}^{p,q}(X) \longrightarrow H^{p,q}(X)$$

is an isomorphism.

Notation.
$$b^m = \dim H^m(X) \quad m(\text{-th Betti number}),$$
$$h^{p,q} = \dim H^{p,q}(X).$$

In the case of a compact Kählerian variety we have

$$b^m = \sum_{p+q=m} h^{p,q},$$

furthermore

$$h^{p,q} = h^{q,p} = h^{n-p,n-q}.$$

The first relation follows from the fact that Δ is a real operator, especially

$$\Delta\omega = 0 \iff \Delta\bar{\omega} = 0,$$

the second relation follows from

$$\Delta\omega = 0 \iff \Delta(*\omega) = 0.$$

XVIII. Differentiable Mappings. If X and Y are two differentiable (resp. analytic) manifolds, one defines in an obvious way the notion of a differentiable (resp. analytic) mapping

$$\varphi : X \longrightarrow Y.$$

If ω is an m-form (resp. (p,q)-form) on Y, one defines the pulled back form $\varphi^*\omega$ on X. We obtain natural mappings

$$\varphi^* : H^p(Y) \longrightarrow H^p(X).$$

They are compatible with the de Rham isomorphism and the mappings

$$H^m(Y, \mathbf{C}) \longrightarrow H^m(X, \mathbf{C})$$

of the singular cohomology groups which are usually defined in algebraic topology.

A similar statement is true in the case of analytic manifolds x, y. In the case of a holomorphic mapping we obtain a commutative diagram

$$\begin{array}{ccc}
\omega & \longmapsto & \varphi^*\omega \\
\\
H^{p,q}(Y) & \longrightarrow & H^{p,q}(X) \\
\| \wr & & \| \wr \\
H^q(Y, \Omega_Y^p) & \longrightarrow & H^q(X, \Omega_X^p).
\end{array}$$

The mapping in the second line is induced by the natural mapping

$$\Omega_Y \longrightarrow f_*\Omega_X .$$

Let now $D \subset \mathbf{R}^n$ ($\mathbf{C}^n$) be an open domain and Γ a group of C^∞-diffeomorphisms (biholomorphic mappings) of D which acts discontinuously and freely (i.e. without fixed points). The quotient space

$$X_\Gamma := D/\Gamma$$

can be equipped with a natural differentiable (analytic) structure such that the projection

$$p : D \longrightarrow X_\Gamma$$

is locally diffeomorphic (biholomorphic). The mapping $\omega \mapsto \varphi^*\omega$ defines isomorphisms

$$M_\infty^p(X_\Gamma) \longrightarrow M_\infty^p(D)^\Gamma$$

$$(M_\infty^{p,q}(X_\Gamma) \longrightarrow M_\infty^{p,q}(D)^\Gamma) .$$

The superscript Γ means that we take the subspace of Γ-invariant elements.

Assume that a Γ-invariant Riemannian (Hermitean) metric on D is given (i.e. all elements of Γ are motions). Then we obtain a natural Riemannian (Hermitean) metric on the quotient X_Γ. The projection

$$D \longrightarrow X_\Gamma$$

is a local isometry. We obtain isomorphisms

$$\mathcal{H}^p(X_\Gamma) = \mathcal{H}^p(D)^\Gamma$$
$$\mathcal{H}^{p,q}(X_\Gamma) = \mathcal{H}^{p,q}(D)^\Gamma .$$

XIX. Poincaré Duality. Let X be a differentiable manifold of dimension n. We define the de Rham cohomology groups

$$H^p(X) = C^p(X)/B^p(X)$$
$$C^p(X) = \text{space of all } C^\infty \text{ } p\text{-forms } \omega , d\omega = 0 ,$$
$$B^p(X) = \text{space of all } d\omega , \omega \text{ a } C^\infty \text{ } (p-1)\text{-form} .$$

One can also define the **de Rham cohomology groups with compact support:**

$$H_c^p(X) = C_c^p(X)/B_c^p(X) ,$$

where

$$C_c^p(X) = \text{space of all } C^\infty \text{ } p\text{-forms } \omega \text{ with compact support} , \quad d\omega = 0 .$$

("Compact support" means of course that there exists a compact subset $K \subset X$ such that the restriction to $X - K$ is zero.)

$$B_c^p(X) = \{d\omega \mid \omega \text{ a } C^\infty \ (p-1)\text{-form with compact support}\}.$$

We have

$$C_c^p(X) \subset C^p(X),$$
$$B_c^p(X) \subset B^p(X),$$

and hence obtain a natural linear mapping

$$H_c^p(X) \longrightarrow H^p(X),$$

which in general is neither injective nor surjective.

The theorem of de Rham states that there are natural isomorphisms

$$H^p(X) \cong H^p(X, \mathbf{C})$$

$$H_c^p(X) \cong H_c^p(X, \mathbf{C})$$

between the de Rham cohomology groups and the singular cohomology groups (with or without compact support) with coefficients in $\mathbf{C}$.

The Poincaré duality theorem is usually proved in the context of singular cohomology. We express it in terms of the "de Rham cohomology".

First we construct a pairing

$$H^p(X) \times H_c^{n-p}(X) \longrightarrow \mathbf{C}.$$

We represent elements of $H^p(X)$ (resp. $H_c^{n-p}(X)$) by differential forms

$$\omega \in C^p(X) \quad (\text{resp. } \omega' \in C_c^{n-p}(X)).$$

We can consider the n-form $\omega \wedge \omega'$. It has compact support and is hence integrable. We claim that the integral

$$\int_X \omega \wedge \omega'$$

depends only on the class of ω or ω'. This means for example that

$$\int_X d\tilde{\omega} \wedge \omega' = 0.$$

We have $d\tilde{\omega} \wedge \omega' = d(\tilde{\omega} \wedge \omega')$ and the assertion follows from a special case of

Stokes's Theorem. *Let ω be a C^∞ $(n-1)$-form with compact support. Then we have*

$$\int_X d\omega = 0.$$

So Stokes's theorem gives us the desired pairing

$$H^p(X) \times H_c^{n-p}(X) \longrightarrow \mathbf{C}.$$

The Poincaré duality theorem states that this pairing is non-degenerate under certain assumptions.

(A bilinear mapping

$$V \times W \longrightarrow \mathbf{C} \quad (a, b) \longmapsto <a, b>$$

for two vector spaces V, W is called non-degenerate if for each $a \in V, a \neq 0$, there exists a $b \in W$ such that $<a, b> \neq 0$ and vice versa. The spaces V and W then have the same dimension.)

We now assume that X is contained as an open subspace in a compact topological space $\overline{X}$. We assume that the topological space

$$\partial X := \overline{X} - X$$

(with the induced topology of $\overline{X}$) is also equipped with a structure as differentiable manifold. We assume furthermore that each "boundary" point $a \in \partial X$ admits an open neighbourhood $U(a)$ and a topological mapping

$$\varphi : U(a) \overset{\sim}{\longrightarrow} V = \{x \in \mathbf{R}^n \mid \|x\| < 1, \; x_n \geq 0\}$$

such that

$$\varphi(U(a) \cap X) = V_0 = \{x \in V \mid x_n > 0\}$$

and such that the mappings

$$U(a) \cap X \longrightarrow V_0,$$
$$U(a) \cap \partial X \longrightarrow \{x \in \mathbf{R}^{n-1} \mid (x, 0) \in V\},$$

induced by φ are diffeomorphisms. (This means that X is the interior of a compact C^∞-manifold with boundary.)

Poincaré Duality. *Under the above assumptions on X we have*

1) All the cohomology groups

$$H^p(X) \quad , \quad H_c^p(X)$$

are finite dimensional.

2) The pairing

$$H^p(X) \times H_c^{n-p}(X) \longrightarrow \mathbf{C}$$

$$(\omega, \omega') \longmapsto \int_X \omega \wedge \omega'$$

is non-degenerate. We especially have

$$\dim H^p(X) = \dim H_c^{n-p}(X).$$

The proof of this theorem is usually reduced to a corresponding result in algebraic topology via the "de Rham isomorphism". But is is also possible to give a proof in the context of differential forms. In this connection we mention another long exact sequence which is well known from algebraic topology.

Under the same assumptions as in the Poincaré duality theorem one has a long exact sequence *

$$\ldots \longrightarrow H_c^p(X) \longrightarrow H^p(X) \longrightarrow H^p(\partial X) \overset{\partial}{\longrightarrow} H_c^{p+1}(X) \longrightarrow \ldots .$$

The nature of ∂ is not important in our application. All other mappings are natural ones.

Example: Let Γ be a discrete subgroup of $SL(2,\mathbf{R})^n$ such that $(\mathbf{H}^n)^*/\Gamma$ is compact. We assume that Γ has no elliptic fixed points. Then the quotient

$$\mathbf{H}^n/\Gamma$$

satisfies the assumptions of the Poincaré duality theorem.

But the compactification by the cusps is not a manifold with boundary. We have to modify this compactification.

Recall that close to the cusp ∞ the quotient $\mathbf{H}^n/\Gamma$ looks like

$$\overline{U}_C/\Gamma_\infty \quad , \quad \overline{U}_C = \{z \in \mathbf{H}^n \mid Ny \geq C\}$$

with $C > 0$. We have a natural topological mapping

$$\overline{U}_C/\Gamma_\infty \overset{\sim}{\longrightarrow} \{t \in \mathbf{R} \mid t \geq C\} \times Y,$$

where

$$Y = \{z \in \mathbf{H}^n \mid Ny = 1\}/\Gamma_\infty.$$

This space carries a natural differentiable structure. We have proved that it is compact. Hence we may compactify $\overline{U}_C/\Gamma_\infty$ by adding not only a single point but by adding $\infty \times Y$.

$$\overline{U}_C/\Gamma \hookrightarrow [C,\infty] \times Y.$$

We may repeat this construction for each cusp class and obtain a realization of $\mathbf{H}^n/\Gamma$ as the interior of a manifold with boundary.

This shows that the Poincaré duality theorem can be applied to $\mathbf{H}^n/\Gamma$.

The spaces

$$\mathbf{H}^n/\Gamma_\infty , \ \overline{U}_C/\Gamma_\infty , \ [C,\infty] \times Y$$

---

* "Exact" means that the image of an arrow equals the kernel of the next one.

are homotopy equivalent. We obtain

$$H^p(\partial X) \cong \bigoplus_{j=1}^h H^p(\mathsf{H}^n/\Gamma_{\kappa_j})$$

where $\kappa_1, \ldots, \kappa_h$ is a set of representatives of the cusps.

We therefore obtain an exact sequence which is used in Chap. III, §5

$$\ldots \longrightarrow H_c^m(\mathsf{H}^n/\Gamma) \longrightarrow H^m(\mathsf{H}^n/\Gamma) \longrightarrow \bigoplus_{j=1}^h H^m(\mathsf{H}^n/\Gamma_{\kappa_j})$$

$$\overset{\partial}{\longrightarrow} H_c^{m+1}(\mathsf{H}^n/\Gamma) \longrightarrow \ldots .$$

All the arrows besides the ∂'s are obvious ones.

Bibliography

Andreotti, A., Vesentini, E.
1. Carleman Estimates for the Laplace-Beltrami equation on complex manifolds. Publ. Math., I.H.E.S. **25**, 313-362 (1965)

Ash, A., Mumford, D., Rapoport, M., Tai, Y.
2. Smooth compactification of locally symmetric varieties. Math. Sci. Press, Brookline, Mass. 1975

Baily, W.L.
3. Satake's compactification of V_n^* . Amer. J. Math. **80**, 348-364 (1980)

Baily, W.L., Borel, A.
4. Compactification of arithmetic quotients of bounded symmetric domains. Ann. of Math. **84**, 442-528 (1966)

Bassendowski, D.
5. Klassifikation Hilbertscher Modulflächen zur symmetrischen Hurwitz-Maaß-Erweiterung. Bonner Math. Schriften **163** (1985)

Blumenthal, O.
6. Über Modulfunktionen von mehreren Veränderlichen. Math. Ann. **56**, 509-548 (1903) and **58**, 497-527 (1904)

Cox, D., Parry, W.
7. Genera of congruence subgroups in **Q**-quaternion algebras. J. f. d. reine u. angew. Math. **351**, 66-112 (1984)

Cartan, H.
8. Fonctions automorphes. Seminaire No. 10, Paris 1957/58

Deligne, P.
9. Théorie de Hodge. I, II Publ. Math., I.H.E.S. **40**, 5-58 (1971) and **44**, 5-77 (1974)

Dennin, J.
10. The genus of subfields of $K(p^n)$. Illinois J. of Math. **18**, 246-264 (1984)

Ehlers, F.
11. Eine Klasse komplexer Mannigfaltigkeiten und die Auflösung einiger isolierter Singularitäten. Math. Ann. **218**, 127-156 (1975)

Freitag, E.
12. Lokale und globale Invarianten der Hilbertschen Modulgruppe. Invent. Math. **17**, 106-134 (1972)
13. Über die Struktur der Funktionenkörper zu hyperabelschen Gruppen. I, II J. f. d. reine u. angew. Math. **247**, 97-117 (1971) and **254**, 1-16 (1972)
14. Eine Bemerkung zur Theorie der Hilbertschen Modulmannigfaltigkeiten hoher Stufe. Math. Zeitschrift. **171**, 27-35 (1980)

Freitag, E., Kiehl, R.
15. Algebraische Eigenschaften der lokalen Ringe in den Spitzen der Hilbertschen Modulgruppen. Invent. Math. **24**, 121-148 (1974)

van der Geer, G.
16. Hilbert modular forms for the field **Q**($\sqrt{6}$). Math. Ann. **233**, 163-179 (1978)

17. Hilbert modular surfaces. Erg. der Math. III/16 Springer-Verlag

van der Geer, G., Zagier, D.

18. The Hilbert modular group for the field $\mathbf{Q}(\sqrt{13})$. Invent. Math. **42**, 93-133 (1977)

Gundlach, K-B.

19. Some new results in the theory of Hilbert's modular group. Contributions to function theory. Tata Institute Bombay165-180 (1960)

20. Die Bestimmung der Funktionen zur Hilbertschen Modulgruppe des Zahlkörpers $\mathbf{Q}(\sqrt{5})$. Math. Ann. **152**, 226-256 (1963)

21. Die Bestimmung der Funktionen zu einigen Hilbertschen Modulgruppen. J. f. d. reine u. angew. Math. **220**, 109-153 (1965)

22. Poincarésche und Eisensteinsche Reihen zur Hilbertschen Modulgruppe. Math. Zeitschrift. **64**, 339-352 (1956)

Hammond, W.

23. The modular groups of Hilbert and Siegel. Amer. J. of Math. **88**, 497-516 (1966)

24. The two actions of the Hilbert modular group. Amer. J. of Math. **99**, 389-392 (1977)

Harder, G.

25. A Gauss-Bonnet formula for discrete arithmetically defined groups. Ann. Sci. E. N. S. **4**, 409-455 (1971)

26. On the cohomology of discrete arithmetically defined groups.

Helling, H.

27. Bestimmung der Kommensurabilitätsklasse der Hilbertschen Modulgruppe. Math. Zeitschrift. **92**, 269-280 (1966)

Hermann C.F.

28. Symmetrische Hilbertsche Modulformen und Modulfunktionen zu $\mathbf{Q}(\sqrt{17})$. Math. Ann. **256**, 191-197 (1981)

29. Thetareihen und modulare Spitzenformen zu den Hilbertschen Modulgruppen reell-quadratischer Körper. Math. Ann. **277**, 327-344 (1987)

Hirzebruch, F.

30. Hilbert modular surfaces. L'Ens. Math. **71**, 183-281 (1973)

31. The Hilbert modular group, resolution of the singularities at the cusps and related problems. Sém. Bourbaki1970/71, exp. 396. In: Lecture Notes in Math. **244**. Springer-Verlag (1971)

32. The Hilbert modular group for the field $\mathbf{Q}(\sqrt{5})$ and the cubic diagonal surface of Clebsch and Klein. Usp. Mat. Nauk **31**, 153-166 (1976) (in Russian) Russian Math. Surveys **31** (5), 96-110 (1976)

33. The ring of Hilbert modular forms for real quadratic fields of small discriminant. In: Modular functions of one variable VI. Lecture Notes in Math. **627**, 287-324. Springer-Verlag (1976)

34. Modulflächen und Modulkurven zur symmetrischen Hilbertschen Modulgruppe. Ann. Sci. E. N. S. **11**, 101-166 (1978)

35. The canonical map for certain Hilbert modular surfaces In: Proc. Chern Symp. 1979. Springer-Verlag (1981)

36. Überlagerungen der projektiven Ebene und Hilbertsche Modulflächen L'Ens. Math. **24**, 63-78 (1978)

Hirzebruch, F., van der Geer, G.

37. Lectures on Hilbert modular surfaces. Les Presses de l'Univ. de Montréal (1981)

Hirzebruch, F., Van de Ven, A.

38. Hilbert modular surfaces and the classification of algebraic surfaces. Invent. Math. **23**, 1-29 (1974)

39. Minimal Hilbert modular surfaces with $p_g = 3$ and $K^2 = 2$. Amer. J. of Math. **101**, 132-148 (1979)

Hirzebruch, F., Zagier, D.
40. Intersection numbers of curves on Hilbert modular surfaces and modular forms of Nebentypus. Invent. Math. **36**, 57-113 (1976)
41. Classification of Hilbert modular surfaces. In: Complex Analysis and Algebraic Geometry. Iwanami Shoten and Cambridge University Press 43-77 (1977)
Knöller, F. W.
42. Zweidimensionale Singularitäten und Differentialformen. Math. Ann. **206**, 205-213 (1973)
43. Ein Beitrag zur Klassifikation der Hilbertschen Modulflächen. Archiv der Math. Vol. XXVI (1975)
44. Elementare Berechnung der Multiplizitäten n-dimensionaler Spitzen. Math. Ann. **225**, 131-143 (1977)
45. Beispiele dreidimensionaler Hilbertscher Mannigfaltigkeiten von allgemeinem Typ. Manuscr. Math. **37**, 135-161 (1982)
46. Über die Plurigeschlechter Hilbertscher Modulmannigfaltigkeiten. Math. Ann. **264**, 413-422 (1983)
Maaß, H.
47. Über Grupen von hyperabelschen Transformationen. Sitzungsber. Heidelb. Akad. Wiss., 3-26 (1940)
48. Über die Erweiterungsfähigkeit der Hilbertschen Modulgruppe. Math. Zeitschrift. **51**, 255-261 (1948)
Matsushima, Y., Shimura, G.
49. On the cohomology groups attached to certain vector valued differential forms on the product of the upper half planes. Ann. of Math. **78**, 417-449 (1963)
Meyer, C.
50. Die Berechnung der Klassenzahl abelscher Körper über quadratischen Zahlkörpern. Berlin (1957)
Mumford, D.
51. Hirzebruch's proportionality in the non-compact case. Invent. Math. **42**, 239-272 (1977)
Prestel, A.
52. Die elliptischen Fixpunkte der Hilbertschen Modulgruppen. Math. Ann. **177**, 181-209 (1968)
53. Die Fixpunkte der symmmetrischen Hilbertschen Modulgruppe zu einem reell-quadratischen Körper mit Primzahldiskriminante. Math. Ann. **200**, 123-139 (1973)
Resnikoff, H.L.
54. On the graded ring of Hilbert modular forms associated with $\mathbb{Q}(\sqrt{5})$. Math. Ann. **208**, 161-170 (1974)
Serre, J.P.
55. Faisceaux algébriques cohérents. Ann. of Math. **61** (1955)
56. Géométrie algébrique et géométrie analytique. Annales de l'Institut Fourier **6**, 1-42 (1956)
Shimizu, H.
57. On discontinuous groups acting on a product of upper half planes. Ann. of Math. **77**, 33-71 (1963)
Siegel, C.L.
58. Lectures on advanced analytic number theory. Tata Institute Bombay. 1961, 1965
59. The volume of the fundamental domain for some infinite groups. Trans. AMS **39**, 209-218 (1936). Correction in: Zur Bestimmung des Fundamentalbereichs der unimodularen Gruppe. Math. Ann. **137**, 427-432 (1959)
Thompson, J.
60. A finiteness theorem for subgroups of $PSl(2,\mathbf{R})$ which are commensurable with $PSl(2,\mathbf{Z})$. Proc. Symp. pure Math. **37**, AMS, Santa Cruz, (1980)

Vaserstein, L.
   61. On the Group $SL(2)$ on Dedekind rings of arithmetic type. Mat. Sbornik **89** (1972)
       (= Math. USSR Sbornik **18**, 321-325 (1972))
Zagier, D.
   62. Modular forms associated to real quadratic fields. Invent. Math. **30**, 1-46 (1975)

Index

E. Freitag, University of Heidelberg;
R. Kiehl, University of Mannheim

Etale Cohomology
and the Weil Conjecture

With a Historical Introduction by J. A. Dieudonné

Translated from the German manuscript by Betty S.
and William C. Waterhouse

1988. XVIII, 317 pp. (Ergebnisse der Mathematik und
ihrer Grenzgebiete, 3. Folge, Vol. 13) Hardcover
DM 188,- ISBN 3-540-12175-7

Contents: Introduction. – The Essentials of Etale
Cohomology Theory. – Rationality of Weil ζ-Func-
tions. – The Monodromy Theory of Lefschetz
Pencils. – Deligne's Proof of the Weil Conjecture. –
Appendices. – Bibliography. – Subject Index.

This book is concerned with one of the most important developments in algebraic geometry during the
last decades. In 1949 André Weil formulated his
famous conjectures about the numbers of solutions of
diophantine equations in finite fields.
He himself proved his conjectures by means of an
algebraic theory of abelian varieties in the one-
variable case. In 1960 appeared the first chapter of
the "Eléments de Géometrie Algébraique" par
A. Grothendieck (en collaboration avec
J. Dieudonné). In these "Eléments" Grothendieck
evolved a new foundation of algebraic geometry with
the declared aim to come to a proof of the Weil
conjectures by means of a new algebraic cohomology
theory.
Deligne succeeded in proving the Weil conjectures on
the basis of Grothendiecks ideas.
The aim of this "Ergebnisbericht" is to develop as
self-contained as possible and as short as possible
Grothendiecks l-adic cohomology theory including
Delignes monodromy theory and to present his original proof of the Weil conjectures.

Springer-Verlag Berlin
Heidelberg New York London
Paris Tokyo Hong Kong

Springer

G. van der Geer, University of Amsterdam

Hilbert Modular Surfaces

1988. IX, 291 pp. 39 figs. (Ergebnisse der
Mathematik und ihrer Grenzgebiete, 3. Folge,
Vol. 16) Hardcover DM 148,-
ISBN 3-540-17601-2

Contents: Introduction. - Notations and
Conventions Concerning Quadratic Number
Fields. - Hilbert's Modular Group. - Resolu-
tion of the Cusp Singularities. - Local Invari-
ants. - Global Invariants. - Modular Curves
on Modular Surfaces. - The Cohomology of
Hilbert Modular Surfaces. - The Classification
of Hilbert Modular Surfaces. - Examples of
Hilbert Modular Surfaces. - Humbert
Surfaces. - Moduli of Abelian Schemes with
Real Multiplication. - The Tate Conjectures
for Hilbert Modular Surfaces. - Tables. -
Bibliography. - List of Notations. - Index.

Over the last 15 years important results have
been achieved in the field of *Hilbert Modular
Varieties.* Though the main emphasis of this
book is on the geometry of Hilbert modular
surfaces, both geometric and arithmetic
aspects are treated. An abundance of examples
- in fact a whole chapter - completes this
competent presentation of the subject. This
"Ergebnisbericht" will soon become an indis-
pensible tool for graduate students and
researchers in this field.

Springer-Verlag Berlin
Heidelberg New York London
Paris Tokyo Hong Kong